COOL ASTRONOMY

First published in the United Kingdom in 2014 by
Portico Books
10 Southcombe Street
London
W14 0RA

An imprint of Anova Books Company Ltd

ISBN 978 1 909396 41 8

A CIP catalogue record for this book is available from the
British Library.

10 9 8 7 6 5 4 3 2 1

Illustrations by Damien Weighill
Printed and bound by Times Offset (M) Sdn Bhd, Malaysia

This book can be ordered direct from the publisher at
www.anovabooks.com

Planet boring!

Take A Deeper Look Inside...

Welcome To Cool Astronomy

Imagine you are a superhero. Imagine you have the ability to look into the skies above you and see everything that, so far, humans have been able to record and observe in the last few thousand years of astronomical development. Even in the clearest and darkest of night skies, and no matter how hard you look or squint, with your superhero eyes, you'll never be able to see any more than 4 per cent of the Universe – and that's if you could see *everything*: every single star, planet and galaxy that exists.

The Universe is just so staggeringly huge that even with the Extremely Large Telescopes that are currently being constructed on Earth, as well as NASA's new James Webb Space Telescope, we astronomers (professional and amateurs alike) will never fully see – and possibly grasp – just how gloriously mammoth the Universe is. But that doesn't mean we should just give up and stop looking. No, we should all grab our telescopes and binoculars and help out!

Astronomy, for me, is like a great detective novel with the last page ripped out. You'll never know whodunit, but the process of looking and discovering and trying to work it all out, or at least putting some of it in some sort of order, is actually the best, and most fun, part.

So, please enjoy *Cool Astronomy*. While it is impossible to squeeze everything – so many billions of years, galaxies and planets (plus billions of stars!) – into just one small book, I have tried to include the bits that I have learned about that tickled me the most. These are the things that made me want to continue learning more about astronomy … and continue looking up in expectation and delight at the stars above. I hope they do the same for you.

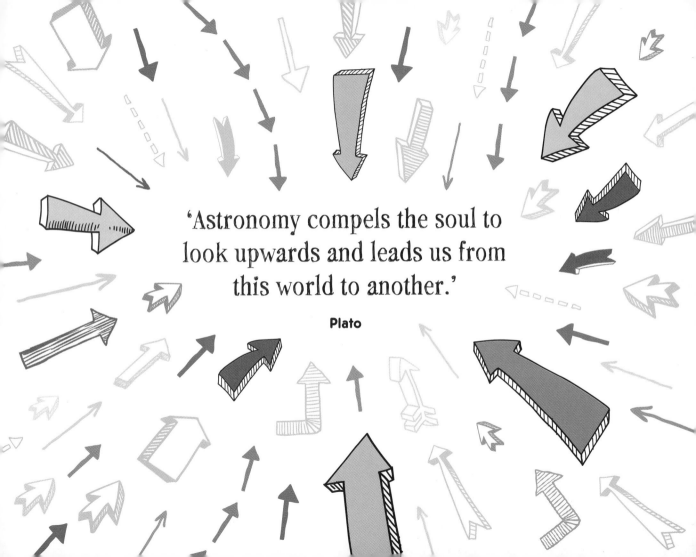

'Astronomy compels the soul to look upwards and leads us from this world to another.'

Plato

The Night Sky – and How to See it

Your eyes are one of the most complex and amazing creations on planet Earth. They are not only the window to your own soul, but also the window to the entire Universe above you. So, get your eyes warmed up (eye-yoga is always fun) and get your peepers 'dark-adapted' and ready for an adventure that is quite literally out of this world …

Look Up!

The skies above us, that we can see with our eyes, and with the help of telescopes both small and Extremely Large, are becoming brighter and better understood. With each new 24-hour cycle of Earth's orbit around its parent Sun, a new star, planet, galaxy or other astronomical phenomenon is being discovered, observed and celebrated by astronomers all over the world. If you want to be the first to discover something new in the sky, then here are the first four things to achieve on your astronomical 'to do' list!

1. Observe the night sky

2. Buy a pair of binoculars or a telescope

3. Learn more about astronomy

4. Join a local astronomy club

Watch This Space

Our Universe is full of wonderful things. And though you'll only ever glimpse a tiny fraction of it through a telescope (less than 1 per cent) that doesn't mean to say you should ever stop looking.

Here is a quick guide to a few amazing things you can see in the sky, with your telescope or binoculars.

Comets Balls of rock and ice with tails of gas and dust.

Stars Balls of gas that sit vast distances away from one another.

Meteors/shooting stars Grains of rock that burn up on entry into Earth's atmosphere.

The Moon A globe of grey rock orbiting our planet.

Planets Spheres of gas or rock, dangling in the night sky.

Galaxies Misty patches of fog or collections of stars, gas and space dust that sit beyond our Milky Way.

Nebulae Giant glowing masses of gas, which eventually form into stars.

The Sun A sphere of hot burning gas, which gives life to everything on Earth.

Milky Way A band of faint light that is a system of stars, gas and dust that makes up our home galaxy. At its heart lies a super-massive black hole.

Auroras Shifting, technicoloured lights, electrically charged (by the Sun), dancing over the polar regions of Earth's atmosphere.

WHAT IS ASTRONOMY?
Astronomy is defined as the branch of science that deals with celestial objects, space, and the physical Universe as a whole.

9

How on Earth do Telescopes Work?

The telescope is one of mankind's greatest astronomical achievements. Which is amazing when you think about it, as all it is is two lenses or mirrors mounted in a tube that makes distant objects seem closer. But when it's pointed towards the stars, the whole universe looks clearer.

Let's Get Started

A telescope is an instrument that collects **electromagnetic radiation** (such as visible light) and makes distant objects seem much closer. The first known 'telescopes' were invented by three Dutch inventors in 1608, but it was an Italian physicist called Galileo Galilei who, in 1609, improved the design and, most importantly, turned it towards the sky.

The Astronomy Bit

There are two types of telescope. A telescope that uses lenses is called a **refracting telescope,** because the lenses **refract** or bend light. The main lens is called the **objective** and additional lenses are used in the eyepiece.

A **reflecting telescope** uses mirrors instead of lenses. Invented by Sir Isaac Newton, a reflecting telescope uses a **concave mirror** instead of the objective lens to form an image, which is then viewed through an eyepiece.

Did You Know?

Due to be complete in 2022, the European Extremely Large Telescope (see page 93) will be located upon the top of the Cerro Armazones mountain in Chile. When built it will be the biggest telescope in the world and will be able to capture clear images of very distant planets. The telescope's integral feature is a main mirror that spans 39 metres (128ft).

REFRACTOR

The simplest refracting telescopes
have two lenses. One convex lens
brings the light from an object towards
a focus, a second lens forms the
eyepiece. Galileo's telescopes had
a single concave lens for an eyepiece,
but an eyepiece can be a complex
construction of several lenses.

REFLECTOR

The simplest reflecting telescope
uses a concave mirror to bring
light from the object towards a
focus. A flat mirror diverts the light
out of the telescope tube, and an
eyepiece is mounted at the side.

Objective lens

Eyepiece lens

REFRACTOR TELESCOPE

LIGHT

NEWTONIAN REFLECTOR TELESCOPE

LIGHT

Secondary mirror

Primary mirror

Bang and Crunch

As we begin to understand astronomy, first we need to know how *everything* in the Universe was created – and ultimately how it all might end. But don't worry, we won't be around to see that!

Look Up!

Most astronomers believe that the entire Universe was created 13.8 billion years ago in a most spectacular fashion that has become known as the Big Bang.

At the time, all of life, everything that we now know to exist, was contained within a bubble a thousand times smaller than the head of a pin. Imagine that! Then one moment – for reasons no astronomer can yet figure out – this bubble exploded and expanded, growing from the size of an atom to the size of a galaxy in a fraction of a second, releasing time, space and matter to create what we now call the Universe.

Approximately 300,000 years after the Big Bang, the Universe started to slow its expansion, cool down and chill out a little bit. When this happened, helium and hydrogen clouds started to appear – the first ingredients in the creation of stars. Once stars formed, the Universe started to take shape …

Watch This Space

Just one of many theories that outlines that ultimate fate of the Universe, the Big Crunch theory was, up until very recently, regarded as how life as we know it would end. As with the Big Bang – the hyper-quick explosion and expansion of everything – the Big Crunch theory related to the ultimate contraction of everything. Instead of an ever-expanding Universe, the Universe would cease stretching (like the pulling of a rubber band) and start collapsing back in on itself, returning to a singularity, or as many theorised, a giant black hole. With the discovery of dark energy and dark matter, modern astronomers believe that the Universe will never be able to stop expanding and forever increase in size. If this is the case, the Big Crunch theory is an impossible ending to life on Earth and everywhere. How do you think the Universe will end? Got any bright ideas?

That's Astronomical!

If you are concerned about the end of the world, then don't worry so much about the Big Crunch. In a few billion years, our Sun will begin burning all of its helium and begin its transformation into a red giant star – en route to becoming a large shell of gas. Once this process begins, and the Sun starts to increase in size rapidly, many astronomers believe that many – if not all – of the planets in our Solar System will be eaten up and consumed.

Gravity

A key element to understanding astronomy is gravity – the force that attracts every object in the Universe towards every other.

Look Up!

Look up at the sky and choose one of our Solar System's eight planets. Take Jupiter, the largest planet in our system (which therefore exerts a greater amount of gravity), for example. Any object in space, such as Jupiter, has a mass, and Newton's law of universal gravitation (1684), as well as Einstein's Theory of General Relativity (1905), outlined that any object that has a mass attracts every object with mass in the Universe. So Jupiter, along with the Moon and the Sun and everything else we know of, is pulling on you and everything else. It is gravity that keeps Earth and the other planets in orbit around the Sun in our Solar System, and it is gravity that causes tides in the seas.

Watch This Space

Gravity has the following effects all around the Universe. Everywhere, in fact:

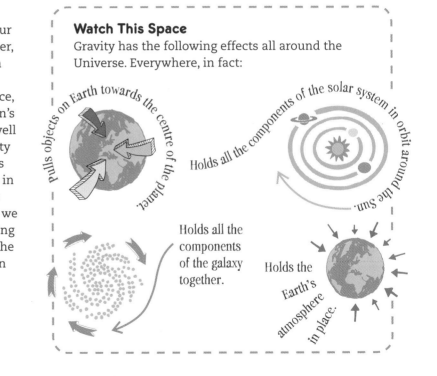

Pulls objects on Earth towards the centre of the planet.

Holds all the components of the solar system in orbit around the Sun.

Holds all the components of the galaxy together.

Holds the Earth's atmosphere in place.

How To Defy Gravity

If gravity is a force that pulls objects towards each other, then what do you think anti-gravity does? That's right, it does the opposite. But it is possible to defy gravity, and here's how. Put a piece of cardboard over the top of a glass of water, making sure the cardboard covers the entire top of the glass. Holding the card in place with your palm, pick up the glass with the other hand and turn it upside down. Then slowly take your hand away from the card. You will find that it does not drop to the floor – instead it will appear to defy all rules of gravity, and disprove that what goes up must come down!

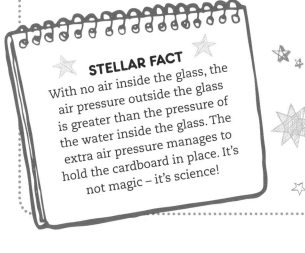

STELLAR FACT

With no air inside the glass, the air pressure outside the glass is greater than the pressure of the water inside the glass. The extra air pressure manages to hold the cardboard in place. It's not magic – it's science!

The Milky Way

While we might think that planet Earth may just be dangling all alone in the middle of the Universe, when viewed from 28,000 light years away Earth is, in fact, sitting cosily in a spiral galaxy we call the Milky Way. Because Earth is located *inside* this galaxy, it means every time you look up, you're looking at *and into* the Milky Way!

Look Up!

The Milky Way, our galactic home, is almost as old as the Universe itself. Containing approximately 200 billion stars (this could be as high as 400 billion!), the Milky Way is known as a spiral galaxy. While nowhere near the largest known galaxy (that prize goes to IC 1101, with over 100 trillion stars), the Milky Way is considered a middleweight galaxy. Full of dust and gas (enough to make many more stars!), the Milky Way has a supermassive black hole at its centre, billions of times larger than the mass of our Sun. Earth, along with the rest of our Solar System, is located around 28,000 light years away from the centre of the Milky Way.

Watch This Space

Our Solar System is located within the Milky Way, a spiral galaxy on a branch of a spiral arm called the Orion Spur. Called the Milky Way because our ancient astronomer experts decided it looked like a thin, hazy river of milk in the sky, the Milky Way is constantly spinning and rotating – remember that next time you feel dizzy! More than half the stars in the Milky Way are older than our Sun, which is 4.5 billion years old.

That's Astronomical!

As late as the 1920s, astronomers thought all of the stars in the Universe were contained inside the Milky Way. It wasn't until American astronomer Edwin Hubble used special stars, known as Cepheid variables, to measure distances precisely that he realized that separate galaxies could be located outside our own.

Build Your Own Space Rocket

The evolution of astronomy in the 20th and 21st centuries has expanded rapidly thanks to space rockets. Without these feats of human scientific engineering we would never have been able to propel satellites and telescopes, such as the legendary Hubble, into orbit to take amazing photographs, nor would we have been able to send astronauts into space, or explore Mars ... and beyond. In tribute to these explosive wonders, lets make our own ... and fire it into space!

Stuff To Get

In order to prepare your rocket for flight, you'll need:

- 1 balloon (round ones will work, but long airship-type ones work best)
- 1 long piece of kite string, about 3–4.5m (10–15ft) long
- 1 plastic straw
- Sticky tape
- Parent's permission
- Pilot's license (just kidding)

Let's Get Started

Once you have everything set up, let's begin our countdown to ignition ...

⭐ **1** Tie one end of the string to a chair.

⭐ **2** Insert the other end of the string through the straw.

⭐ **3** Pull the string tight and tie it to another support in the room, such as a doorknob.

⭐ **4** Blow up the balloon (but don't tie it).

⭐ **5** Pinch the end of the balloon and tape the balloon to the straw as shown in the illustration.

⭐ **6** It's launch time ... let go of the balloon whenever you're ready!

Straw

Sticky tape

String

> T MINUS THREE, TWO, ONE ...
> **THRUST**!

Balloon

That's Astronomical!
So how does this air-propelled rocket work? As the air rushes out of the balloon, it creates a forward motion called thrust. Thrust is a pushing force created by energy. In this case thrust comes from the energy of the balloon forcing the air out. In a real space rocket, thrust is created by the force of burning rocket fuel as it blasts from the rocket's engine – as the engines blast down, the rocket goes up.

Stellar Fact

The Chinese invented and developed the first rockets around 1200 AD. One thousand years later, NASA's most recent, and thrilling, Deep Space 1 project is based on xenon ion engines, which thrust electrically charged particles called ions, not hot gases, out of the back of the spacecraft. It is the first ever use of this type of rocket propulsion and represents the future of flinging objects into space.

19

The Solar System

With our Sun accounting for 99.9 per cent of all the mass in the Solar System (90 per cent of the remaining 0.1 per cent is taken up by Jupiter and Saturn), and with its size approximately 1,392,045km (865,000 miles) across, it's no wonder that the group of planets that Earth belongs to in our Milky Way is called the Solar System.

Look Out!

Our Solar System was born 4.5 billion years ago, when a dark, cold cloud of hydrogen mixed with other gases and 8 billion-year-old space dust. This cloud – many light years in size – collapsed under its own gravitational force and began to spin, and in doing so became a massive, rotating disc. While spinning, the space dust began to collide and clump together, creating denser clumps that grew into large chunks of rock and metal.

THE SUN 1,392,045km (865,000 miles) in diameter

Watch This Space

The planets in our Solar System are divided into two types – gas giants and terrestrial, rocky planets.

After 50 million years the Sun turned into a star, ignited by a thermonuclear reaction of immense proportions. As the solid clumps of rock and metal continued to collide, they eventually formed rocky planets – Mercury, Venus, Earth and Mars – in the inner region of the Solar System.

The four giant gas planets within our Solar System (Jupiter, Saturn, Uranus and Neptune) were formed in the cooler outer regions of the Solar System, when chunks of rock and ice gathered around supermassive volumes of gas.

Pluto
(dwarf planet)

The Eight Planets

MERCURY 57.9 million km (36 million miles) from the Sun

VENUS 107.8 million km (67 million miles) from the Sun

EARTH 149.7 million km (93 million miles) from the Sun

MARS 227.9 million km (142 million miles) from the Sun

JUPITER 777.3 million km (483 million miles) from the Sun

SATURN 1,429 million km (888 million miles) from the Sun

URANUS 2,871 million km (1,784 million miles) from the Sun

NEPTUNE 4,496 million km (2,794 million miles) from the Sun

That's Astronomical!

Think you've discovered a planet? According to the 2006 Ruling on Planets, there are three rules a celestial body must comply with to be classified as a planet.

⭐ **1** It has to orbit the Sun.
⭐ **2** It has to have enough gravity to pull itself into a sphere.
⭐ **3** It needs to have cleared out its orbit of other objects.

Create Your Own Solar System!

If you are interested in looking at the stars, planets and galaxies that are scattered across the Universe, then why not make your own Solar System in your back garden. You'll not only get a feeling of how big our own 'cosmic neighbourhood' is, but also have some fun in the process!

Look Out!

The eight planets in our Solar System display a variety of sizes, colours, ingredients and features. They are not, in any shape or form, similar to each other. For example, over 1,300 Earths could fit inside Jupiter and 1.2 million Earths (or 900 Jupiters) could fit snugly inside the Sun. The sizes and chemical makeup of each planet varies because of the way the Solar System was born 4.5 billion years ago.

Stellar Fact

There is a vast cloud of frozen comets called the Oort Cloud, which surrounds our Solar System. It lies roughly 50,000 times further from the Sun than the Earth. How many more squares of toilet paper would you need to put the cloud on our model?

Watch This Space

To get a sense of how big the Solar System is, we could do a terribly long and boring exercise and work it out on a calculator ... or we could just use some toilet paper!

Stuff To Get
- Toilet roll with over 200 sheets (a standard roll is 240 sheets)
- A felt-tip pen
- Sticky tape (for repairs!)
- Permission to use the toilet roll (ask mum)
- A quiet street, park or large garden – you'll need a distance of about 26m (85ft)

Let's Get Started

Use this table to mark (or draw) the correct planet on each appropriate piece of toilet roll. Then, when you have rolled out the toilet paper to the correct length, measure the corresponding number of sheets away and illustrate the next planet on the list. Use a felt-tip pen, or print out drawings of the planets themselves. Start with the Sun on sheet one and roll outwards from there, and you'll soon get a sense of the sheer scale of the Solar System ...

PLANET	SQUARES OF TOILET ROLL FROM THE SUN	SHEET DIFFERENCE
Mercury	2	—
Venus	3.7	1.7
Earth	5.1	1.4
Mars	7.7	2.6
Ceres (asteroid)	14	6.3
Jupiter	26.4	12.4
Saturn	48.4	22
Uranus	97.3	48.9
Neptune	152.5	55.2
Pluto (dwarf planet)	200	47.5

The Nearest Stars

There may well be 200 billion stars in our galaxy, so there's plenty for you to discover! Can you name the nearest star to Earth? Ask a friend and see if they know the answer. Many people will try to guess names of stars they might know, like Alpha Centauri, or Betelgeuse, but they'd be wrong. The answer, obviously, is our Sun.

Look Up!

To become a budding astronomer, a knowledge of celestial bodies, such as nearby stars and prominent constellations close to Earth, is super-important. Not only do these local stars act as a map of the sky, helping us locate and pinpoint other harder-to-see objects, but they also give us a clue as to the forms of other star types that may be found too.

The 20 Nearest Stars to Earth

STAR	DISTANCE (LIGHT YEARS)	STAR	DISTANCE (LIGHT YEARS)
Alpha Centauri	4.3	Ross 128	10.9
Barnard's Star	5.9	EZ Aquarii	11.3
Wolf 359	7.8	Procyon	11.4
Lalande 21185	8.3	61 Cygni	11.4
Sirius	8.6	Struve 2398	11.5
Luyten 726-8	8.7	Groombridge 34	11.6
Ross 154	9.7	Epsilon Indi	11.8
Ross 248	10.3	DX Cancri	11.8
Epsilon Eridani	10.5	Tau Ceti	11.9
Lacaille 9352	10.7	GJ 106	11.9

Watch This Space

Though it's hard to calculate, astronomers believe that once your eyes are 'dark-adapted' (i.e. you've been looking up at the sky for longer than 20 minutes) you could be treated to the sight of around 5,000 stars visible to the naked eye.

For your astronomical purposes, let's learn about Alpha Centauri – the closest star (its actually a 'triple star system') to our Sun.

That's Astronomical!

According to NASA, there are 45 stars within 17 light years of the Sun, but Sirius is the brightest star in our night sky, due not only to its size but also its proximity to us. Have you spotted it?

THE SUN

ALPHA CENTAURI A

Alpha Centauri B

Proxima

An easy one to spot – and never stare at it!

A yellow star, like our Sun.

25 per cent greater diameter than our Sun.

Third brightest star seen from Earth (behind Sirius and Canopus).

Most North Americans never see it.

Visible about 4.2 light years from the Sun.

Alpha Centauri is the brightest star in the southern constellation Centaurus the Centaur.

Alpha Centauri's surface temperature is a few degrees Kelvin less than our Sun (that is, about 5,770K).

Strange Planets

The Universe is full of strange and alien locations we'd all like to visit on holiday, even if it would take forever to get there and we'd perish the very instant we set foot on it. You might not be able, yet, to spot these planets through your telescope as they are so far away, but keep on looking ...

Look Up!
The first planet discovered outside our Solar System, in 1992, was celebrated around the world. Now, over 20 years later, more than 1,000 exoplanets have been confirmed beyond our Solar System – and some of them really are quite strange.

DIAMOND-IN-THE-ROUGH PLANET

Diamond Planet – or 55 Cancri e – is known as Super Earth, a rocky planet orbiting a Sun-like star.

❋ Twice the radius of Earth.

❋ A mass eight times greater than Earth's.

❋ 40 light years from Earth.

❋ One year around its sun is 18 hours long.

❋ ⅓ of the planet (the weight of three Earths!) is covered in diamond!

❋ Surface level temperature is 2,149°C (3,900°F)!

❋ It is composed mainly of carbon crushed down to form a thick layer of diamond with a molten-iron core.

LAVA PLANET

In October 2013, a puzzling Lava Planet – real name Kepler-78b – was discovered. An Earth-like twin in size and composition, its orbit pattern currently baffles astronomers.

- 20 per cent wider than Earth.
- 80 per cent bigger than Earth.
- A mass 1.69 times greater than Earth's.
- Orbits a Sun-like star in the constellation Cygnus, 400 light years from Earth.
- Completes a year of its parent star every 8.5 hours (one of fastest alien planets ever detected).
- Surface level temperature is 2,027°C (3,680°F) – hot enough to melt iron!
- This alien world should not exist where it does – this planet formed inside a star!
- The planet's days are numbered – in three billion years or so.
- 100 times closer to its parent star, than Earth is to our Sun.

That's Astronomical!

A team of astronomers estimated in 2013 that every Milky Way star (around 200 billion) has, on average, 1.6 worlds. This means that our galaxy potentially has 160 billion planets in it!

Stellar Facts

Planets discovered located outside our own Solar System are called extrasolar planets, or exoplanets. The largest exoplanet ever discovered, so far, is also one of the strangest and theoretically should not even exist, scientists say. Named TrES-4, the planet is about 1.7 times the size of Jupiter.

Einstein and the Speed of Light

In astronomy light, in all its various wavelengths, is very important.
In fact, it's the most important thing. If stars, galaxies or planets did not
produce any light (of any wavelength) then not only would we not know
they were there, we would also not be able to calculate how far away
they are, and we'd have nothing beautiful to look at every night.

> ' ... we establish by definition that the "time"
> required by light to travel from A to B equals the
> "time" it requires the light to travel from B to A.'
> **Albert Einstein**

299 792 458

299,792,458m
(983,571,056ft)
per second

Look Up!

There is nothing faster in the Universe than the speed of light. Which is helpful,
because it's faster than anything we'll ever need on Earth. The first person to
almost-accurately measure the speed of light was Dutchman Olaf Roemer,
in 1676, but it was Albert Einstein who, in the early part of the 20th century,
devised his Special Theory of Relativity and calculated that the speed of light
is always constant: no matter what you do to it, the speed of light will always
remain the same. It was in this theory he formulated his famous equation,
$E = mc^2$, or energy equals mass multiplied by the speed of light squared.

Watch This Space

In his theory Einstein stated that it is impossible to determine whether or not you are moving unless you can look at another object, and he theorised that all movement is relative to other objects. For example, at the moment you are reading this book and you are not moving at all, but relative (i.e. from their perspective) to the distant galaxies, stars and planets in space, *you are moving at nearly the speed of light*. Relative to the Earth, most meteorites move at about 25,000 miles an hour, but if you were standing on a meteorite looking at another meteorite going in the same direction as you and at the same speed, it would not appear to move at all.

$$E = mc^2$$

1 light year = 9.4605284×10^{15} metres or 9,460,528,400,000,000 metres (5,878,499,811,100 miles)

That's Astronomical!

In 2011, the Oscillation Project with Emulsion Tracking Apparatus, or OPERA experiment, at the Large Hadron Collider at CERN, Geneva hit the headlines all around the world when they observed neutrinos – tiny subatomic particles – appearing to travel faster than light. The world went crazy, believing that scientists had discovered a new particle that could break Einstein's law of Special Relativity. But it was not to be. In the end the computer calculating vast amounts of data during the experiments had got the numbers wrong!

STELLAR FACT

Light takes 1.255 seconds to get from the Earth to the Moon and, approximately, 8 minutes and 20 seconds from the Sun to the Earth. So when you look at the Sun (not for too long) you're seeing it as it was 8 minutes and 20 seconds ago.

With These Rings, I Thee Wed: Saturn

Saturn is one of the most beautiful planets to observe through a telescope. Even though Saturn's ring system is not unique – gas giants Neptune, Jupiter and Uranus also have rings – Saturn's famous rings are the most distinct. But how much do you know about them?

Look Up!

Described by Galileo as an object 'with ears', Saturn – the sixth furthest planet from the Sun in our Solar System – is a gas giant made up of 90 per cent hydrogen and 10 per cent helium. Through a telescope it glows dark yellow and Saturn's distinct rings have made it an astronomer's favourite. The six bright, unique and broad ring strands consist of orbiting ice particles, some as small as a grain of sand, others as large as mountains; Saturn's satellite moons are imperative to the creation and existence of its rings.

That's Astronomical!

Each ring travels around Saturn at a different speed and the rings produce rain that falls onto the planet.

Watch This Space

Formed of thousands of ringlets and billions of particles, as captured so amazingly by the Cassini spacecraft, Saturn's rings are classified alphabetically (in the order they were discovered), to distinguish them from one another. They are believed to be pieces of asteroids, comets and shattered moons that broke up before they reached the planet. With a thickness of about 1km (3,281ft) or less, they span up to 282,000km (175,227 miles), about three-quarters of the distance between the Earth and its Moon. The spaces between the rings are created by the orbit pattern of many of the moons circling the planet.

D ring
Distance 66,970 (41,613 miles) – 74,490km (46,286 miles)
Width 7,500km (10.874 miles)
The innermost ring, composed mainly of dust.

Distances are measured from the planet centre to the start of the ring.

C Ring
Distance 74,490km (46,286 miles)– 91,980km (57,154 miles)
Width 17,500km (10.874 miles)
This ring has been dubbed 'Crepe Ring' because, well, it looks like a big crepe!

B Ring
Distance 91,980km (57,154 miles)– 117,580km (73,061 miles)
Width 25,500km (15,845 miles)
The broadest and brightest of the rings.

A Ring
Distance 122,050km (75,838 miles)–136,770km (84,985 miles)
Width 14,600km (9.072 miles)

F Ring
Distance 140,224km (87,131 miles)
Width 30km (18½ miles)–500km (311 miles)
This ring is held together by two moons, Prometheus and Pandora.

G Ring
Distance 166,000km (103,148 miles)–174,000km (108,119 miles)
Width 8,000km (4,971 miles)

E Ring
Distance 180,000km (111,847 miles)–480,000km (298,258 miles)
Width 300,000km (186,411 miles)
The vast outer ring, composed mainly of fine ice grains.

SATURN FACTS
Moons 62; Titan is the largest.

Average distance from Sun 1,434 million km (891 million miles).

Length of day 10.656 hours.

Length of year 29.4 Earth years.

Atmosphere Hydrogen 97 per cent, helium 3 per cent.

Hands in the Sky

Being an astronomer doesn't always mean you have to have fancy, and expensive, equipment. Sometimes your best instruments are the ones right in front of your face!

Zenith point this way!

1°

Look Up!

As the old saying goes, 'I know such-and-such as well as the back of my hand'. Well, now you'll know space like the back of your hands – an essential skill for finding your way around the sky.

Watch This Space

When looking up at the sky you will see 180° of vision. Astronomers use degrees to measure large distances and sizes in the sky. For example, 90° is the distance from the horizon (the point directly in front of you) to the zenith point (the point directly above your head).

When you look through a telescope, your field of vision is minimized to just one degree. At this stage, astronomers use what are called arc minutes and arc seconds to measure distances. There are 60 arc minutes in 1° and 60 arc seconds in one arc minute.

Horizon

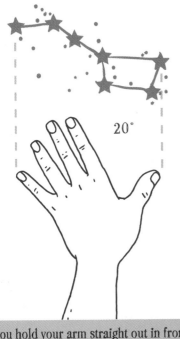

20°

The highest point in the sky is called the zenith.

10°

2°

If you hold your arm straight out in front with the back of your hand facing you, the distance from the tip of your thumb to the tip of your little finger is 20° – roughly the distance between the first and last stars of the Big Dipper asterism.

You can measure smaller distances with your clenched fist at arm's length: the width of your fist is equivalent to 10°.

Your little finger should cover the entire Moon and represents 1°, while your thumb is 2°. After becoming familiar with this technique, you should develop a feel for the distances between the various stars or planets.

Earth's Place in Space

You'd think that weighing 6,600,000,000,000,000,000,000 tonnes would cause anything to fall from the sky, but no: after five billion years Earth is still in one piece. Let's take a minute to reflect on the beauty of our planet's astronomical place in space.

Look Up!

If you have been fortunate enough to travel on a plane, you'll know that looking down on Earth from 10,671m (35,000ft) is a truly wonderful thing – you get to see our planet from the same perspective as the stars. Imagine then the view of the astronauts and cosmonauts currently floating around in the International Space Station 402km (250 miles) above sea level.

Watch This Space

In 1990, the Voyager I space probe viewed our planet from a distance of 6 billion km (3.7 billion miles) and sent the image back to Earth. This photo, called The Pale Blue Dot, is one of the most famous pictures ever taken, not because you see Earth in all its majesty, but because it is seen as nothing more than a tiny dot against the vastness of space.

Stellar Fact

The largest meteor ever found on Earth left no crater – the Hoba meteorite was flat on both sides so may have skipped through our atmosphere, like a stone skimming on water.

International Space Station

Earth is the only planet in our Solar System with tectonic plates – without them Earth would overheat!

The Earth's rotation is gradually slowing 17 milliseconds per hundred years, which means in 140 million years time our ancestors will have 25 hours in their day!

The temperature of Earth's core is as hot as the surface of the Sun.

Each day is 23 hours 56 minutes and 4 seconds long – not 24!

106 billion people have existed on planet Earth to date.

Earth is 4.54 billion years old!

38,000 man-made objects have orbited the Earth since Sputnik I in 1957.

100 tonnes of small meteorites enter the Earth's atmosphere every day.

That's Astronomical!
There are records dating back to 1600 AD of a Chinese space enthusiast, called Wan Hoo, who attempted to launch himself into space by fixing 47 gunpowder rockets to his body – and then ordering his servants to light them! I don't think he ever made it, but at least he went out with a bang.

Earth's nickel-iron core and speedy rotation create a powerful magnetic field. This field protects the Earth from the effects of harmful winds sent from the Sun.

Stars of Astronomy: Part I

Since Man started looking up at the stars and realizing how cool they were – nearly 5,000 years ago – there have been a select group of astronomy heroes who, by opening their minds, have expanded our knowledge and allowed us to glimpse back to the very creation of the Universe.

Galileo Galilei

➡️ *Born* Pisa, Italy, 1564

⭐ Galileo dramatically improved the telescopes of his time – increasing magnification from × 3 to × 30!

⭐ Galileo confirmed that the Sun is the centre of our Solar System, a theory first put forward by Nicolaus Copernicus. This was revolutionary at the time because most of the world still believed Earth was the centre of the Universe.

⭐ First to observe that the Moon was cratered and that Jupiter had four moons.

⭐ First to observe that there were thousands of stars invisible to the naked eye.

⭐ Galileo wrote in simple, concise language that everyone could understand.

⭐ He also observed that there were 'spots' on the Sun.

⭐ Because Saturn is tilted, when its rings are facing Earth edge-on, they 'disappear' from our view. We now know this happens every 14 years or so, but poor Galileo questioned his sanity when they 'disappeared' and then 'reappeared' a few years later.

Isaac Newton

→ **Born** *Lincolnshire, England, 1643*

★ Newton invented the reflecting telescope in 1688 – the most common type of telescope still used today.

★ He discovered the laws of gravity and established the Three Universal Laws of Motion in 1687. These laws are:

1. An object will remain at rest or moving in a straight line unless acted upon by an external force.

2. When force is applied to an object, it will accelerate (Force = mass × acceleration).

3. For every action, there is an equal and opposite reaction.

★ Newton invented calculus.

★ In 1696, Newton recalled all English coins, had them melted down and turned into ones that were harder to counterfeit. This left the country without currency for an entire year, but effectively established a proper Mint.

> Force = mass × acceleration

Albert Einstein

→ **Born** *Ulm, Germany 1879*

★ 'Albert Einstein' is an anagram of 'Ten elite brains'.

★ In his 'miracle' year of 1905, Einstein discovered the Special Theory of Relativity, the $E = mc^2$ equation (which posited that all matter can be turned into energy), and the idea of the quantum – the physical phenomenon of matter at subatomic levels.

★ Einstein developed his Special Theory of Relativity in five weeks. It took him four years to develop the General Theory of Relativity.

> $E = mc^2$

Stars of Astronomy: Part II

Three more heroes of astronomy that you'll need to know about before you start your quest to explore the stars.

Johannes Kepler

→ **Born** *Weil der Stadt, Germany, 1571*

- The father of celestial mechanics.
- Kepler coined the words 'satellite' and 'orbit'.
- He was the first to suggest that the Sun rotates about its axis.
- Kepler devised the Three Laws of Planetary Motion, which state:

  1. The path of the planets about the Sun is elliptical in shape, with the centre of the Sun being located at one focus (the Law of Ellipses).

  2. An imaginary line drawn from the centre of the Sun to the centre of the planet will sweep out equal areas in equal intervals of time (the Law of Equal Areas).

  3. The ratio of the squares of the periods of any two planets is equal to the ratio of the cubes of their average distances from the Sun (the Law of Harmonies).

- Kepler formulated eyeglass designs for near- and far-sightedness.

Nicolaus Copernicus

➤ **Born** *Torun, Poland, 1473*

⭐ The father of modern astronomy.

⭐ Copernicus's *On The Revolution of the Celestial Spheres*, changed the way we view the cosmos.

⭐ He invented the idea of the heliocentric solar system – that the Sun was the centre of the Universe. This shattered the geocentric theory, believed since Roman times, 2,000 years earlier. Imagine waking up one day and discovering that the Earth was no longer the centre of the Universe!

⭐ Copernicus was the first scientist to suggest the Earth was not a stationary object.

⭐ He posited the idea that the Sun rotates on its axis once every day.

⭐ His name is not pronounced 'Copper Knickers.'

Edwin Hubble

➤ **Born** *Missouri, USA, 1889*

⭐ A pioneer of extragalactic astronomy.

⭐ Hubble made two important astronomical discoveries:

1. In 1923, that our galaxy is not the only one.

2. In 1929, that the farther a galaxy is from Earth, the faster it appears to move away. This notion of an expanding Universe formed the basis of the Big Bang theory – that the Universe began with an intense explosion of energy at a single moment in time and has been expanding ever since.

⭐ In the 1930s, Hubble claimed that the galaxies are evenly distributed in space. This was a mistake, and other scientists were quick to disprove it. Whoops!

Connect The Dots

An ordinary pair of eyes can see about 5,000 stars in the night sky. Astronomers call these naked-eye stars. But, as you won't be able to name them all (just yet) there is another way to view, and remember, these stars easily – not as individuals, but as groupings known as constellations.

Look Up!

A constellation is a group of stars that, when viewed from Earth, form a pattern or shape in the night sky. In 1925, the International Astronomical Union adopted 88 official constellations and assigned specific constellation names to areas of the sky. Although many of the 88 constellations don't resemble the creatures or characters they are named after, don't worry – from an astronomical point of view, constellations are just a useful way to help identify positions of stars in the sky.

That's Astronomical!

The brightest constellation visible in the night sky is Crux (or the Southern Cross). The constellation with the greatest number of visible stars is Centaurus (the Centaur) with 101 stars. The largest constellation is Hydra (the Water Snake), which extends over 3.158 per cent of the sky!

Watch This Space

It's easy to forget, when connecting the dots within constellations in the sky, that while individual stars within a constellation may appear to be very close to one another, in fact they can be separated by huge distances and have no real connection to each other whatsoever.

Top 5 Major Constellations visible above the horizon

1. Ursa Minor
2. Ursa Major
3. Cassiopeia
4. Cepheus
5. Draco

Signs of the Zodiac

Can you name all 12 Zodiacal constellations and their corresponding animals? Look away and test yourself!

♍ Virgo the Virgin

♎ Libra the Scales

♑ Capricorn the Goat

♌ Leo the Lion

♐ Sagittarius the Archer

♒ Aquarius the Water Bearer

♈ Aries the Ram

♊ Gemini the Twins

♓ Pisces the Fish

♉ Taurus the Bull

♋ Cancer the Crab

♏ Scorpius the Scorpion

How to Become a NASA Astronaut

At some stage everybody wants to be an astronaut when they grow up. But have you got what it takes? If you want to get a close look at the stars then you may need to put your games console away and get cracking – it takes a lot of hard work. NASA only accepts the best of the best!

Look Up!

The NASA Astronaut Program in the USA is one of the toughest job interview processes in the world, but it's not impossible. With the Space Shuttle out of commission since 2012, the only means of travelling in space is in rockets, but with the burgeoning space tourism program initiated by Virgin Galactic, and the forthcoming Mars One project, the need for space pilots in 2014 (and beyond!) is at an all time high.

Watch This Space

NASA selects astronauts from a diverse pool of applicants with a wide variety of backgrounds. From the thousands of applications NASA receives each year, only a few are chosen for the intensive Astronaut Candidate training program, at the Johnson Space Center, in Houston, Texas. Only 330 astronauts have been selected to date.

The basic requirements for an Astronaut Pilot, include the following:

1. A bachelor's degree from an accredited institution in engineering, biological science, physical science or mathematics. An advanced degree is desirable.

2. At least 1,000 hours pilot-in-command time in jet aircraft. Test-flight experience is a necessity.

3. Ability to pass a NASA space physical which is similar to a military or civilian flight physical and includes the following specific standards:
   ★ **Distant visual acuity** 20/100 or better uncorrected, correctable to 20/20 each eye.
   ★ **Blood pressure** 140/90 measured in a sitting position.
   ★ **Height** between 157.5cm (5ft 2in) and 190.5cm (6ft 3in).

That's Astronomical!
The word 'astronaut' derives from the Greek words meaning 'space sailor', a wonderfully illustrative word that conjures up images of heroic men sailing through an ocean of stars in wonderful space boats!

Stellar Facts

All food eaten in space is precooked or processed so it requires no refrigeration.

You can't boil pasta in space – the bubbles in the pan don't rise!

Space Shuttle astronaut Bill Thornton once opened a packet of M&Ms for a bedtime snack. When some of the candy floated away, they later returned – while he was asleep – and smacked him in the face!

Here Comes the Sun

The Sun is the centre of Earth's affections. It provides us with warmth, life, and as a constant feature in our sky, it isn't very difficult to locate. Many astronomers believe that the Sun is about halfway through its life: it's five billion years old, so there are only another five billion more to go!

Look Up!

At the Sun's extremely hot core, 620 million tonnes of hydrogen are converted into helium every second. When the Sun's life began, it changed from a protostar to a truestar when thermonuclear reactions started to occur at its core. The Sun is one big nuclear reactor, generating vast amounts of energy – if all of the Sun's energy could be collected it could power the entire USA for the next nine million years!

Stuff To Get

To sketch the Sun, you'll need:
- Binoculars
- A small mirror
- A pad of paper
- A drawing pencil
- A room with a window, facing the Sun.

That's Astronomical!

In around 5 billion years from now the Sun will have become a red giant star, and by doing so may engulf all the planets in the Solar System. But for now, the Sun sits 149.6 million km (93 million miles) away.

Corona the Sun's intensely hot outer atmosphere.

Chromosphere a layer of glowing red hydrogen gas.

Photosphere the visible surface of the Sun.

Flare an explosion of hot gas that shoots waves of charged particles out across the Solar System towards Earth.

Let's Get Started

You should never look directly at the Sun – the Sun's harmful and powerful rays can cause blindness, so please be careful. Because of this, astronomers have devised a handy way of observing the Sun, and in particular its famous 'spots' – without looking directly at it.

⭐ **1** Place a telescope or pair of binoculars (with lenses no bigger than 50mm/2cm) in an open window in the direct path of the Sun's rays. Make a white cardboard collar (30cm/12in square)to fit around the lens.
⭐ **2** Next, place a small mirror in front of one eyepiece of the binoculars so that it throws an image of the Sun on a wall of the room. Adjust the mirror until the image is sharp and darken the room.
⭐ **3** Now you can view the Sun and its spots cool, highly magnetized areas on the Sun's surface – on the wall.

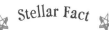

⭐ Stellar Fact ⭐

A solar eclipse occurs when the Moon passes in front of the Sun and blocks out most of the Sun's light. The Sun's 1,391,000km (864,000 mile) diameter is 400 times greater than that our Moon (3,476km/2,160 miles). But the Moon also happens to be about 400 times closer to the Earth than the Sun, so the Moon can appear to completely cover the Sun! How amazing is that?

Planets in Motion

When looking up at the sky, everything may look very still. But nothing could be further from the truth. Everything we see in the sky is moving very quickly, not only away from each other, but also in orbit.

Look Up!
Our Earth is orbiting, in an ellipse, around the Sun. Because the Sun's mass is so large, it has attracted all of the planets in the Solar System towards it due to gravity, and now they move in a beautifully orchestrated dance around it.

Each planet in the Solar System has its own elliptical orbit around the Sun. From an astronomical point of view, each planet's orbit is remarkable – the Sun is pulling everything towards it, but because the planets in the Solar System have their own velocity, they have been caught in a spin around the Sun while never crash-landing into it.

The Sun has less of a pull on the planets in the outer Solar System so they revolve around the Sun in a slower orbit. Since they revolve more slowly there is less centrifugal force (where an object pushes away from another object, the opposite of gravity), but also less gravity since they're further away, so they remain in a stable orbit. The planets closer to the Sun orbit much faster, as they are closer and their centrifugal force is greater, neatly counteracting the greater gravity of the Sun.

The Orbits of the Planets

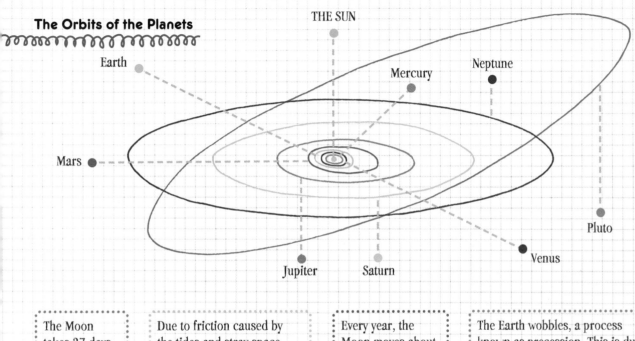

THE SUN

Earth

Mercury

Neptune

Mars

Pluto

Venus

Jupiter   Saturn

The Moon takes 27 days, 7 hours, 43 minutes and 11 seconds to orbit the Earth.

Due to friction caused by the tides and stray space particles, the Earth's rotation is very gradually slowing down, which means longer days! At the beginning of the Earth's formation, a day lasted for 13 or 14 hours.

Every year, the Moon moves about 4cm (1½ in) out from its orbit around the Earth.

The Earth wobbles, a process known as precession. This is due mainly to outside gravitational forces pulling on the Earth's equatorial bulge. The Sun and Moon pull on Earth's bulge, making the planet wobble!

The Asteroid Belt

Earth is surrounded, quite literally, by asteroids. If you were an astronaut and wanted to travel to Jupiter, on your way you would have to pass through one of astronomy's most fascinating pieces of fashion – the Asteroid Belt.

Look Up!

The Asteroid Belt contains asteroids that are leftovers from the formation of our Solar System. During this time, the birth of Jupiter prevented any other planets from forming in the huge gap between Mars and Jupiter, causing the small objects that were there to collide with one another and fragment into the asteroids that are seen today.

Asteroids are too small to be called planets, but are still pretty big. Astronomers have observed that the mass of all the asteroids in the Belt could fit into our Moon!

Watch This Space

The Asteroid Belt, located in a vast ring between Mars and Jupiter, contains over 200 asteroids larger than 96.5km (60 miles) in diameter (they have their own moons!), as well as over 750,000 asteroids that are around 1km (²/₃ mile) in size and millions of smaller ones – thought to be anywhere from the size of a city to just 6m (20ft).

That's Astronomical!

Astronomers have estimated that an asteroid capable of causing a global disaster would have to be more than 400m (¼ mile) wide. Such an impact would produce enough dust in the atmosphere to initiate a 'nuclear winter.' NASA believes asteroids that large strike Earth only once every 1,000 *centuries* on average. However, smaller asteroids are believed to hit Earth every 1,000–10,000 years and could destroy a city or cause devastating tsunamis.

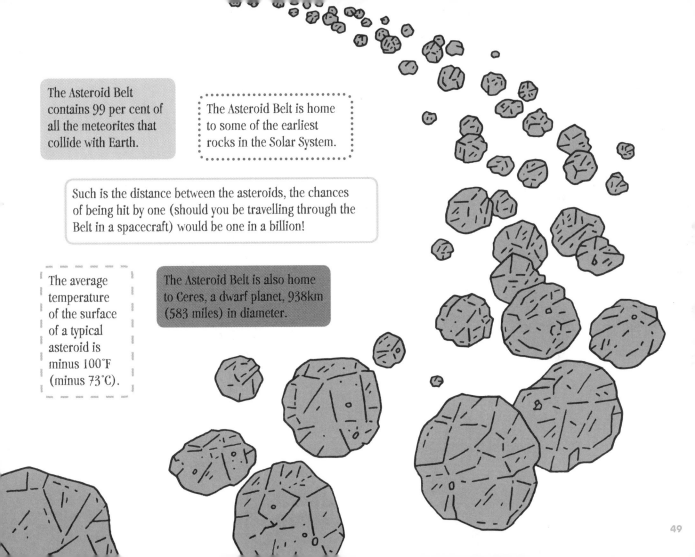

The Asteroid Belt contains 99 per cent of all the meteorites that collide with Earth.

The Asteroid Belt is home to some of the earliest rocks in the Solar System.

Such is the distance between the asteroids, the chances of being hit by one (should you be travelling through the Belt in a spacecraft) would be one in a billion!

The average temperature of the surface of a typical asteroid is minus 100°F (minus 73°C).

The Asteroid Belt is also home to Ceres, a dwarf planet, 938km (583 miles) in diameter.

Keep an Eye Out

Once you have mastered the basics of astronomy, grab your telescope and go hunting for these cosmic stunners! Some are harder to spot than others ... but practice makes perfect.

Look Up!

Our cosmic neighbourhood – as well as beyond – is a massive playground for astronomers. If all the humans that have ever existed all discovered stars, we would not even have discovered all the stars in our own galaxy – let alone the trillions of other galaxies out there. But don't let that get you down. Instead, point your telescope upwards and keep looking.

What To Look Out For

Here are five 'local' wonders to whet your astronomical appetite and keep you busy as you begin your stargazing hunt. Some are trickier to find than others.

1. THE MOON'S TERMINATOR

The Moon is a great place to start. The Lunar Terminator is the Moon's equivalent of the division between night and day. When sunlight strikes this section of the Moon between the light and the dark sides, shadows cast by craters and other geological features are lengthened, making many of the Moon's features easier to see.

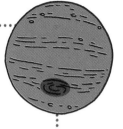

2. JUPITER'S GREAT RED SPOT
Jupiter is the largest planet in the Solar System and at certain times of the year you can pick out in detail the Great Red Spot, a hurricane larger than Earth that has been raging on the planet's surface for centuries.

3. ANDROMEDA GALAXY
One of the nearest major galaxies you can try to find is the Andromeda Galaxy, 2.5 million light years away, located in the constellation Andromeda. Just beyond Andromeda, see if you can spot the satellite galaxies M32 and M110.

4. M15
Found in the constellation Pegasus, M15 is one of the most amazing clusters of stars in the sky that isn't difficult to spot. If you have a small telescope you will only see it as a fuzzy blob, so you'll need a larger one to reveal the individual stars within. But a fuzzy blob is a good start!

5. ORION NEBULA
The Orion Nebula is a colourful jewel in the sky. Located just below the three-star belt in the constellation of Orion, the nebula is a monster stellar wonder. It can be seen with the naked eye, but binoculars or a small telescope will show it in its full glory.

Looking Back in Time

When you look into space you are looking back in time – the light you see is actually light that has travelled vast distances and through time to appear before your eyes. And because distances in space are so incredibly large, the measurements we use on Earth just aren't up to the job.

Look Up!

Astronomers do not measure cosmic distances in kilometers or miles, they use light years. One light year is the distance light travels in one year at the speed of light (see page 28).

Astronomers reckon that in one light year light can travel about 9.5 trillion km (6 trillion miles). So, if you go outside into your garden and flash a torch into the sky, in 1 light year that light will have travelled 9,500,000,000,000km (6,000,000,000,000 miles)!

Watch This Space

Let's create an example. So, if one light year = 9.5 trillion km (6 trillion miles), how far away is Proxima Centauri, the nearest star to Earth, outside our Solar System, if it is 4.3 light years away?

Easy!

9.5 trillion km (6 trillion miles) × 4.3 = 40.85 trillion km (25.8 trillion miles).

And because a light year is the time it takes for light to reach our eyes from the object in space we are viewing, when we look through our telescopes at Proxima Centauri, we don't see it as it is now, but as it was it 4.3 years ago.

That's Astronomical! Travelling by spacecraft, it would take us 81,000 years to reach Proxima Centauri.

First galaxies

First stars

Dark age

Cosmic microwave radiation

The James Webb Space Telescope has been in planning since 1996. It is designed to study the birth and evolution of galaxies by observing the furthest objects in the Universe and thereby looking back in time. If it is completed it will view the Universe from an orbit around the Sun but 1,500,00km (930,000 miles) beyond the Earth.

James Webb Space Telescope

0
Big Bang!

.0004

.3

.95

13.8
Modern Universe

AGE OF THE UNIVERSE (BILLIONS OF YEARS)

The First Sky Watchers

Thousands of years ago, when people first looked to the stars, the first sky watchers believed they were seeing religious deities looking down on them. They might still be right ...

Look Up!

The Babylonians, a civilization located in present-day Iraq almost 4,500 years ago, were the first people we know of who kept astronomical records. In fact, the origins of modern astronomy can be found in the recordings of Mesopotamia, and all modern efforts in the exact sciences descend from the work of the late Babylonian astronomers.

Following the Babylonians, the Chinese, Egyptians and the Greeks all created their own version of early calendars: simple recordings of the movement of the Sun and Moon, and of what we now know as time. The first calendar ever discovered, by the Chinese, was from around 1300 BC.

Stellar Fact

☆

The Crab Nebula was produced by a supernova explosion in 1054 AD; at the time Chinese astronomers noted that the explosion was so bright that it was visible during the day, and lit up the night sky for months!

Watch This Space

Even though they were ultimately proved wrong, here are three ancient Greek philosophers who were at least thinking along the right lines.

THALES (624–547 BC)

Thales was a philosopher who brought astronomical records from Mesopotamia and Egypt back to Greece. Thales believed that the Earth was a disc floating on an endless ocean.

HIPPARCHUS (190–120 BC)

Widely considered to be the greatest astronomer of ancient times, Hipparchus compiled the first known star catalogue of celestial objects. He also estimated the distance from the Earth to the Moon to be '29.5 Earth diameters' (the real value is 30 Earth diameters – so he got it almost right!). Hipparchus' greatest revelation was discovering Earth's wobble caused by the gravitational pull of the Sun and Moon.

ARISTOTLE (384–322 BC)

A great philosopher, Aristotle wrote *De Cealo et Mundo* ('On the Heavens and Earth'), one of the earliest known works of astronomy. He proved that the Earth is a sphere and claimed that it was at the centre of the Universe. According to Aristotle, the Sun, planets and all the stars were located in spheres that revolved around the Earth. Close, but no cigar.

Astronomical Achievements

Just as the sky is littered with stars, the history of the Universe – and astronomy's development through time over the past few thousand years – is littered with tiny, average, massive and supermassive achievements. Let's shine a spotlight on them.

13.8 billion years ago The Big Bang.

3,000 BC The main stones of Stonehenge are laid down.

2000 BC The first solar lunar calendars are devised in Egypt and Mesopotamia.

280 BC Greek philosopher Aristarchus suggests the Earth revolves around the Sun. He also provides the first estimate of the distance from the Earth to the Sun.

240 BC Greek philosopher Eratosthenes measures the circumference of the globe with surprising accuracy.

130 BC Hipparchus develops the first accurate star list with over 850 of the brightest stars.

45 BC The Julian Calendar, a purely solar calendar, is introduced to the Roman Empire.

140 AD Greek philosopher Ptolemy, in his famous work, *Mathematike Syntaxis*, introduces the geocentric theory that everything revolves around the Earth.

813 AD Al Mamon founds the Baghdad School of Astronomy.

1054 Chinese astronomers observe a supernova in Taurus.

1259 In Iran, the world's first observatory is built in honour of the famous Persian astronomer Nasir al-Din al-Tusi.

1543 Nicolaus Copernicus publishes his heliocentric model of the Universe and changes the world!

1572 Revered astronomer Tycho Brahe discovers a supernova in Cassiopeia.

1603 German astronomer Johann Bayer introduces the Bayer designation of stars still in use today. This is the system of assigning Greek letters to the brightest stars.

1608 Dutch spectacle maker Hans Lippershey invents the telescope.

1609 Galileo improves Lippershey's telescope and discovers the moons of Jupiter and the Moon's craters.

1609 Johannes Kepler introduces his Three Laws of Planetary Motion in his work, *Harmonices Mundi* ('Harmony of the World').

1656 Dutch astronomer Christian Huygens discovers Saturn's rings.

1668 Isaac Newton invents the world's first reflecting telescope.

1675 Danish astronomer Olaf Roemer measures the speed of light.

1675 Frenchman G D Cassini observes a split in Saturn's rings and calls it the Cassini Division.

1687 Newton publishes his *Philosophiae Naturalis Principia Mathematica*; modern astronomy as we know it is born.

1705 Englishman Edmond Halley predicts the return of Halley's Comet in 1758; he was right.

1781 Uranus is discovered by Englishman William Herschel.

1843 German astronomer Samuel Heinrich Schwabe discovers the sunspot cycle.

The Goldilocks Zone

On Earth it was just the right amount of heat, light, water and air that caused life blissfully to burst forth. The same rule applies in space – welcome to the Goldilocks Zone.

Look Up!

Earth is what is known as a Goldilocks Planet – it falls into a star's circumstellar habitable zone (CHZ) – a region around a star where planets such as Earth can support liquid water on their surface; a place far enough away from danger to be able to support life with a non-threatening and stable atmospheric climate, but not so far away to be removed from a light and heat source – ingredients vital to life. It is neither too hot, nor too cold.

Watch This Space

Astronomers looking for life on exoplanets in our own Milky Way (where there could be as many as 100 billion planets) first look for signs of habitable zones – or Goldilocks Zones – as this is potentially where life could thrive most successfully.

'This porridge is too hot,' Goldilocks exclaimed.

So she tasted the porridge from the second bowl.

'This porridge is too cold.'

So she tasted the last bowl of porridge.

'This porridge is just right!' she said happily.

And she ate it all up.

Life is highly improbable, except around Sun-like stars – a mere 5 per cent of all stars in the Universe.

MARS – too cold; too far away from the Sun to support life.

EARTH – just right.

VENUS – too hot; too close to the Sun to accommodate life.

Astronomers reported that there could be as many as 40 billion Earth-sized planets orbiting in the habitable zones of Sun-like stars and red dwarf stars just within the Milky Way.

The Universe Within Us All: Atoms

While we humans may feel alone in the Universe, let's take comfort in the knowledge that every single thing ever created is all made up of the same stuff – atoms.

Electron

Proton

Hydrogen is the first element in the Periodic Table, which means it has an atomic number of 1, or 1 proton in each hydrogen atom.

In 1776, hydrogen was recognized as a unique substance by English scientist Henry Cavendish.

It is the lightest element in the Universe as well as the most abundant. About 75 per cent of the element mass of the Universe is hydrogen.

Look Up!

Atoms are the basic chemical building blocks of all matter in the Universe – every star, planet and galaxy you can see through a telescope is made up of atoms. Even the telescope itself is made up of uniquely combined atoms, called molecules, and materials that are made out of many molecules, known as polymers. So, for example, when one atom of oxygen joins with two atoms of hydrogen, we get water (H_2O).

That's Astronomical!

A cube of sugar contains as many atoms as there are stars in the Universe.

Watch This Space

Atoms are made up of protons (that carry a positive electric charge), neutrons (that carry no electric charge) and electrons (that carry a negative electric charge). Because electrons carry a negative electric charge, and protons carry a positive charge, the attraction between them holds the electrons together.

How to Make Your Own Crater!

If we ever see an asteroid heading towards Earth we'll probably be too busy screaming and running away, but what we should really be doing is stopping to take photos, as it is such a rare event.

Look Up!

The world went meteoroid crazy in February 2013 when a 17m (56ft) chunk of space rock weighing 10,000 tonnes burned up in Earth's atmosphere over Russia, travelling at around 64,372km/h (40,000 mph). It was in the atmosphere for 32.5 seconds before exploding as a massive fireball around 19–24km (12–15 miles) above the Earth's surface – creating a violent shock wave that blew out thousands of windows and caused walls to collapse. NASA stated that the explosion was equivalent to 300,000 tonnes of dynamite and more powerful than the atomic bomb that destroyed Hiroshima, Japan in 1945. NASA expects this type of event to happen every 100 years.

STELLAR FACTS
⭐ According to a study, a meteoroid will hit a human being about once every 180 years.
⭐ A 30.5cm (12in) thick coating of Kevlar protects the International Space Station from meteoroids.

That's Astronomical!

Around 500 meteoroids are believed to reach the Earth's surface every year, but only five or six of those are ever recovered for scientists to study.

Let's Get Started

If this meteoroid had landed on Earth it would have made a fairly impressive crater. The Earth's surface, as well as the Moon's, is littered with crater sites. Why don't we create our own craters – it's simple!

First find a sand tray or large, shallow plastic box and fill with sand, then locate each of the objects listed in the box below. Next find a safe place to drop them from, preferably at different heights (use a stepladder!), and record the difference in craters the objects make, in the table provided. Then smooth out the sand ready for the next impact.

You'll see first hand the different types, depths and shapes created by meteoroid craters – the same that impact on the Earth every day. Try it with different-sized objects such as a tennis ball, a ping-pong ball and an apple, and measure the difference in size of the impact each time.

	HEIGHT	CRATER SIZE (cm/in)
Football		
Tennis ball		
Ping-pong ball		
Large potato		
Apple		
Marble		

The Space Between

We now know that if an alien is looking at Earth from a planet 65 million light years away, it will not see any humans, it will see dinosaurs roaming around! But what about the space between the planets? Astronomers need to worry about the distances between objects as well as the objects themselves.

Look Up!

Because the Universe is such an UNBELIEVABLY SUPER-MONSTER MASSIVE place to live in, if we ever wanted to leave Earth (and one day we may have to), it would take us entire lifetimes to reach other destinations even in our own back yard, the Milky Way, let alone beyond those in other 'nearby' galaxies that are still many light years away.

That's Astronomical!

The speed of light is so fast that a beam of light can travel seven times around the Earth in one second.

Stellar Fact

Black holes are so dense, and produce such intense gravity, that even light cannot escape!

Watch This Space

It takes 8.3 minutes for the light from the Sun to reach Earth, but what about light from other 'local' stars and planets? As astronomers, we must learn about light and know the distances of celestial objects from Earth.

DISTANCE	TIME
Earth to the Moon	1.25 seconds
Earth to the Sun	8.3 minutes
Sun to Jupiter	41 minutes
Sun to Saturn	85 minutes
Sun to Neptune (furthest planet from Sun)	4.2 hours
Sun to Voyager I (furthest man-made object from Earth)	17.1 hours
Sun to Alpha Centauri (nearest star system to Earth)	4.3 years
Sun to Sirius (brightest star in our sky)	8. 6 years
Sun to 61 Cygni (binary star)	11. 4 years
Sun to Polaris (the North Star)	432 years
Sun to Orion Nebula (brightest nebula to naked eye)	1,300 years
Sun to Galactic Centre (centre of the Milky Way)	27,700 years
Sun to Andromeda Galaxy	2,540,000 years
Edge of observable Universe	47,000,000,000 years

Hubble, Hubble, Hubble

There are scores of space-based telescopes in a geo-stationary orbit above the Earth – each one carrying out a particular purpose, photographing the Universe using their own special wavelength technology. But it is Hubble, the world's greatest space-based telescope, which will always mean the most to astronomers.

Look Up!

A joint project between NASA and the European Space Agency, the Hubble Space Telescope (named after revered American Astronomer Edwin Hubble) was launched on 25 April 1990 from the Space Shuttle Discovery.

Stellar Fact

☆

Hubble transmits 120 gigabytes of data (not just images) back to Earth every week. That's equal to about 1,097m (3,600ft) of books on a shelf!

Watch This Space

At a staggering 13.7m (45ft) in length, weighing over 11,340kg (25,000lb), located in orbit 568km (353 miles) above Earth and powered by solar panels, Hubble travels around 8km (5 miles) per second. Due to its elevated position away from Earth's atmosphere, Hubble has taken iconic images of some of the remotest galaxies, as well as making discoveries such as the Pillars of Creation of the Eagle Nebula, 1995 – probably the most famous astronomical image of the 20th century.

That's Astronomical!

In order to take photos of distant objects, Hubble must be extremely steady and accurate. The telescope is able to lock onto a target, let's say Jupiter, without deviating more than the width of a human hair when seen at a distance of 1.6km (1 mile)!

When Hubble was first switched on in its new place in space, its engineers on Earth discovered it had a flawed mirror, and the first images it sent back were fuzzy. The telescope's main mirror had a severe defect caused by a manufacturing error. It was three years before NASA could send a repair mission. In December 1993, the first new images from Hubble reached Earth. They changed the world.

Hubble taught us that nearly all galaxies are anchored by supermassive black holes.

Thanks to the Hubble Space Telescope, we now know that the Universe is 13.8 billion years old.

The Hubble Space Telescope has helped scientists determine the process of how planets are born.

The Hubble Space Telescope detected the first organic molecule to be discovered on a planet outside our Solar System.

Hubble has helped determine the rate of the Universe's expansion.

Earth-like Planets

Astronomers have recently calculated that one in every five of the billions of Sun-like stars in the Milky Way has at least one terrestrial planet like Earth orbiting in the Goldilocks Zone (see page 58). So, let's go on the hunt for some Earth-like planets ...

Look Up!

In 2013, NASA's Kepler Space Telescope (launched in 2009), which was built solely to hunt and locate new planets, came back with some startling evidence: there are about 20 billion Earth-like planets in the Milky Way alone, and all of them are located in the Habitable Zone, where life could flourish. According to the new data, this means there could be an Earth-like planet just 12 light years away – its parent star may even be visible to the naked eye.

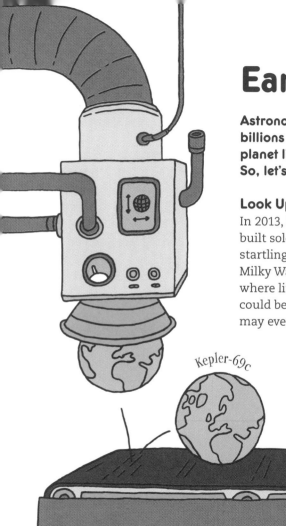

Kepler-69c   Kepler-62e   Kepler-62f

Let's Rename The Planets!

Bolstered by the news of the Kepler Space Telescope's discoveries, let's have a look at three Earth-like planets that are currently blowing astronomers' minds.

Astronomers used to be very imaginative when naming planets, but this doesn't seem to be the case any more. If you could rename these Earth-like planets, either after yourself or a character, what would you call them?

1. Kepler-69c

A Venus-like planet 70 per cent larger than Earth 2,700 light years away.

2. Kepler-62e

A water world in the Goldilocks Zone 1,200 light years away.

3. Kepler-62f

A planet that might suit us, only it's 1,200 light years away.

That's Astronomical

Discovered in 2009, an exoplanet (see page 110) called GJ1214b that is six times bigger than Earth, has been discovered orbiting a red dwarf star 40 light years away. Referred to as a 'steamy waterworld', the planet is made up of 75 per cent water and ice and 25 per cent rock.

STELLAR FACT
It isn't just planets that could be habitable. Moons could suit life too, including Europa, one of Jupiter's moons.

Dear Moon ... Back Soon, Love Earth

The history and existence of Earth is tied very closely to its nearest cosmic neighbour, the Moon. No longer thought to be made of cheese, and with most of modern astronomy focused on looking as far away from Earth as possible, our grey friend in the sky is now also one of astronomy's most overlooked wonders.

Look Up!

Our poor Moon has had a difficult life. As detailed in the Giant Impact Hypothesis (also called the Big Splash), the Moon was formed when an object the size of Mars struck the Earth, flinging masses of debris into orbit around our planet that eventually, over millions of years, clumped and melted together, and then cooled down. As if that wasn't chaotic enough, for another 500 million years, the Moon was left scarred by thousands of meteors – rocks that remained after the Giant Impact – when the Moon was forming. Without the Moon, some biologists believe there would be no life on Earth.

Moon Fact Checklist

As an astronomer, how much do you know about our nearest cosmic friend?

○ The Moon is an average of 384,403km (238,857 miles) away. Its elliptical course means that it gets as close as 363,104km (225,622 miles) away.

○ We only ever see one side of the Moon, because its rotation takes exactly the same amount of time as its orbit of Earth – a spooky celestial coincidence.

○ The Moon's mass is 73,476,730,924,573,500kg (161,988,463,180,000,000lb); it is the fifth largest moon in the Solar System.

○ As it weighs less than the Earth, the Moon has much weaker gravity; so you, for example, would weigh about one-sixth (16.5 per cent) of what you do on Earth.

○ The Moon has no atmosphere.

○ In December 2013 China's Jade Rabbit Moon rover made the first 'soft' landing on the Moon since 1976. Its mission is to test new technologies, gather scientific data and build intellectual expertise.

That's Astronomical!

In 1958, the US Air Force developed a top-secret plan, code-named Project A119, to detonate a nuclear bomb on the Moon. Though it was never carried out, the purpose of the project was to demonstrate a show of force after falling behind the Soviet Union in the Space Race. Any nuclear explosions on the Moon would have been visible to the naked eye on Earth – a clear sign by the USA that they meant business.

The Scale of the Universe

Not only is the Universe already big, it is also accelerating and getting bigger. But in order to appreciate just how astonishingly massive it is – leading astronomers to believe that human brains will never understand just how big it is – we also need to understand the concept of scale. Let's look at the smallest and biggest things in the Universe, and from there we should be able to gasp in wonder at the enormousness of the Universe.

Look Up!

If we take what is currently known to be the smallest object in the Universe, a quark – a tiny subatomic particle that makes up protons and neutrons (these, along with electrons, are the three ingredients that make up atoms) – then what is the biggest thing? In order to be able to appreciate just how big stars and planets are, first we need to appreciate how small the things that create them are.

A QUARK A super-small, subatomic particle that carries a fractional electric charge. Quarks have not been directly observed but theoretical predictions based on their existence have been confirmed experimentally.

Watch This Space

Size is important in astronomy, but so is distance. The most distant galaxy ever seen in the Universe and the current record holder, is the galaxy MACS0647-JD, which is about 13.3 billion light years away. As the Universe itself is only 13.8 billion years old, this means that MACS0647-JD's light has been travelling towards Earth for almost the whole history of space and time.

HYDROGEN ATOM The most abundant particle found in the Universe, almost 75 per cent! A single positively charged proton, and a single negatively charged electron, orbiting its nucleus.

A COFFEE BEAN The key ingredient in what makes coffee taste and smell so good. Coffee and coffee shops are the most profitable, and fastest-growing business on Earth – worth $100 billion dollars worldwide: the world drinks over 500 billion cups a year!

THE SOMBRERO GALAXY 29 million light years away. Like our Milky Way, Sombrero is a spiral galaxy that is easily seen through an amateur telescope. The galaxy is 50,000 light years across – or half the size of our own Milky Way.

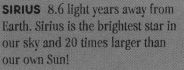

SIRIUS 8.6 light years away from Earth. Sirius is the brightest star in our sky and 20 times larger than our own Sun!

A BLUE WHALE At 30.5m (100ft) long and weighing 170 tonnes, this is the largest creature ever to have lived on Earth. Their hearts can weigh as much as a car, and their tongues alone can weigh as much as an elephant!

That's Astronomical!

The apparent brightness of any object in the sky is due to its intrinsic brightness and its distance. For example, the brightest known object in the Universe is a very faint quasar – 3C 273 – as seen from the Earth, and can only be seen with a large telescope. If it were as close as the nearest galaxy, it would be easily visible to the naked eye. Discovered in 1959, 3C 273 was the first ever quasar to be identified (see page 107).

The Universe's Largest Things

The feeling you get when you spot something mega through your telescope is amazing.
As we know size is relative – a cat is bigger than a mouse, but not as big as a planet!
Let's take a look at some of the Universe's largest objects for you to keep an eye out for.
They'll be easier to find through a telescope than a mouse, at least!

Look Up!

Earth may seem big to us, but it is small potatoes when compared to our celestial neighbour, the gas giant Jupiter (into which the Earth could fit over 1,300 times); and in turn Jupiter is tiny when compared to WASP-17b, the largest planet discovered so far with a radius twice the size of Jupiter. For astronomers, nothing in space is certain, but so far, there have been some truly HUGE discoveries ...

Watch This Space

Grab your telescope and keep a look out for these objects, the top five LARGEST things in your Universe – but just because they are big, relative to our size, that doesn't mean they are easy to spot!

LARGEST ASTEROID – PALLAS
Pallas is among the largest, and most visible, asteroids discovered so far, and accounts for a whopping 9 per cent of the entire mass of the Asteroid Belt.

LARGEST STAR – NML CYGNI
In 2012, NML Cygni was verified as one of the Universe's largest known stars. It would take a beam of light 6 hours and 40 minutes to circle it once; a beam of light can go around the Earth 7.5 times in one second.

LARGEST BLACK HOLE – NGC1277
The biggest black hole yet discovered, located 250 million light years from Earth, in the constellation Pegasus, with a mass equivalent to 17 billion of our Suns and over 14 per cent of its host galaxy's overall mass – most take up just 0.1 per cent of their galaxy's overall mass.

That's Astronomical!
Most astronomers agree that the largest 'thing' in the Universe is the Cosmic Web. Try to imagine a 3-D spider's web of endless galaxy clusters surrounded by dark matter. The Cosmic Web is the skeleton that holds all the galaxies together and dates back as far as 7 billion light years.

LARGEST PLANET – WASP-17B
WASP-17b is the largest planet discovered so far. Most planets move around their parent stars in the same direction as their star moves but WASP–17b moves in the other direction – a retrograde orbit – befuddling astronomers all over the world.

LARGEST GALAXY – IC1101
Almost 1.07 billion light years from Earth IC1101, in the constellation Virgo, is the largest galaxy discovered so far. Spotted in 1790, it has a diameter of 6 million light years (60 times larger than the Milky Way) and is home to roughly 100 trillion stars!

One Giant Leap

The history of astronomy dates back to around 5,000 years ago. But the most ground-breaking developments for modern astronomers have been in the last century or so.

1897 Discovery of the electron.

1905 Einstein's Special Theory of Relativity is published.

1915 Discovery of Proxima Centauri, the nearest star to the Earth (apart from the Sun).

1916 Publication of Einstein's General Theory of Relativity, which predicts the expanding Universe.

1923 Edwin Hubble proves that galaxies are systems independent of the Milky Way.

1929 Hubble presents evidence for the expansion of the Universe.

1930 Discovery of Pluto by Clyde Tombaugh.

1931 Karl Jansky discovers radio waves from space.

1937 Grote Reber reveals radio waves coming from the Milky Way, using his radio telescope.

1938 Theory of stellar energy using nuclear reactions announced by Hans Bethe.

1942 James Hey detects radio waves from the Sun.

1948 Herman Bondi and Thomas Gold propose the steady-state cosmological theory (now-defunct), the alternative to the Big Bang.

1948 George Gamow and Ralph Alpher describe the origin of elements in the Big Bang.

1961 Yuri Gagarin makes first manned space flight.

1951 First space flight by living creatures when the USA sends four monkeys into the stratosphere.

1957 Sputnik 1 launched – the beginning of the Space Age.

1958 NASA founded.

1959 First pictures of the Moon's far side taken by space probes.

1962 John Glenn becomes first American to orbit Earth in space.

1963 Maarten Schmidt identifies quasars for the very first time.

1963 Valentina Tereshkova becomes first woman in space in Vostok 6.

1965 Soviet astronaut Alexei Leonov makes the first space walk; the Russians launch Salyut I, the first orbital space station.

1965 Discovery of the cosmic microwave background radiation by Arno Penzias and Robert Wilson; they win the Nobel Prize for Physics.

1971 First detailed close-range pictures of Mars sent back by Mariner 9.

1973 Pioneer 10 does the first fly-by of Jupiter.

1969 Apollo 11 Moon landing. Neil Armstrong and Buzz Aldrin take the first steps on the lunar surface.

1976 Space probes Viking 1 and 2 land on Mars.

1971 Discovery of the rings of Uranus.

1977 Launch of Voyager I.

1979–81 The Voyager spacecraft pass Jupiter and Saturn, relaying an enormous amount of information back.

1981 Shuttle Columbia is launched.

1983 Pioneer 10 is the first spacecraft to achieve escape velocity from the Solar System.

1983 Infrared Astronomical Satellite (IRAS) completes the first full survey of the infrared sky.

1985 Halley's Comet returns, as correctly predicted by Edward Halley. It will return again on 28 July 2061.

1987 Supernova SN1987A flares up, becoming the first supernova visible to the naked eye since 1604.

1989 Voyager 2 reaches Neptune, discovering a ring system and 8 moons.

1994 An asteroid passes Earth at only 161,000km (100,000 miles).

1990 Hubble Space Telescope is launched.

1990 First exoplanet discovered.

1994 Comet Shoemaker-Levy 9 collides spectacularly with Jupiter.

2006 Pluto is reclassified as a dwarf planet.

1998 Discovery of dark energy is made by Saul Perlmutter and Brian Schmidt.

1998 Construction begins on the International Space Station.

2006 Dark matter is confirmed.

1996 NASA scientists wrongly announce proof of living organisms on a Mars meteoroid in Antarctica.

2012 Curiosity, a Mars rover, lands successfully on the surface of Mars.

Satellites and Space Junk

In the 1950s, satellites were top-secret military objects with hush-hush assignments. Nowadays, they are essential for helping the people of the world to communicate with one another, tracking and predicting weather, watching TV and, of course, carrying out scientific experiments and observations in space. Without satellites our world would come to a halt.

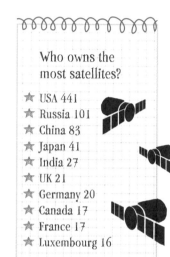

Who owns the most satellites?

★ USA 441
★ Russia 101
★ China 83
★ Japan 41
★ India 27
★ UK 21
★ Germany 20
★ Canada 17
★ France 17
★ Luxembourg 16

Look Up!

Since 1957 – the launch of the first satellite Sputnik I – over 7,000 satellites have been launched into space. Satellites can come in many shapes and sizes, from low-Earth orbiting to geostationary, from the size of a beach ball to the size of a car. Low-Earth satellites operate at heights between 161km (100 miles) and 1,931km (1,200 miles) above the Earth's surface. A geostationary satellite travels from west to east over the equator and moves in the same direction and at the same rate Earth is spinning.

That's Astronomical!

NASA believes there are 23,000 pieces of space junk – objects large enough to track through radar – hurtling through space very quickly, some as large as tennis balls, others as small as marbles.

Watch This Space

Earth is a satellite. Any object that orbits another object can be referred to as a satellite.

Of all the thousands of satellites sent into space, many have different functions and purposes. The most common types of satellite are:

Weather Satellites that help meteorologists predict the weather.

Communications Satellites that allow telephone conversations around the world to be relayed through the satellite. The most important feature of this type of satellite is the transponder, a radio that receives a conversation at one frequency and then amplifies it and retransmits it back to Earth on another frequency.

Navigational Satellites that help ships, planes and cars on Earth navigate. They are called GPS – Global Positioning Satellites.

Military A lot of what military satellites do remains secret, but could cover intelligence-gathering, reconnaissance, encrypted communication, nuclear monitoring, observing enemy movements and early warning detection of missile launches.

Broadcast Satellites that broadcast television signals from one point to another (similar to communications satellites).

Scientific The Hubble Space Telescope is the most famous scientific satellite, but there are many others looking at everything from sunspots to gamma rays.

Earth observation Satellites that observe the planet, from temperature to ozone, climate change to typhoons and hurricanes.

The Electromagnetic Spectrum

Light, microwaves, radio waves, ultraviolet, gamma radiation, X-ray and infrared – these are all very important to astronomers. When we look at an image of a star or galaxy, often these are composites of several different colours overlaid on top of one another. Using different sources of light to look at planets and stars gives astronomers clues about a star's chemical make-up, size, distance and temperature.

Look Up!

When you think of light, you probably think of what your eyes can see. But the light to which our eyes are sensitive is just a tiny portion of the total amount of light that actually surrounds us. Most of the light in the Universe is invisible to our eyes.

In astronomy the electromagnetic spectrum describes all the wavelengths of light and has many practical uses: gamma radiation, the shortest wavelength and highest frequency of light, helps doctors kill cancer cells through radiotherapy; X-rays famously can take pictures of our bones; ultraviolet radiation can tan the skin via sunbeds; visible light is what we see with our eyes; infrared light is what is used in fibre-optic communications (for the Internet etc.); microwaves are great for cooking; while radio waves – the longest and lowest wavelength of light – are used to broadcast television and radio signals.

RADIO WAVES
Radio, television, mobile phones, wireless routers

MICROWAVES
Microwaves

The electromagnetic spectrum is a term devised by scientists to describe the entire range of light that exists in the Universe.

Watch This Space

Try to imagine light as waves crossing an ocean. Like waves, light has a few fundamental properties that describe it. The first is frequency – measured in hertz (Hz) – this counts the number of waves that pass a point in one second. Think of a wave hitting the beach, followed closely by another one. The other property of light is wavelength – this is the distance from the peak of one wave to the peak of the next.

Frequency and wavelength are linked: the higher the frequency, the shorter the wavelength!

INFRARED RADIATION
Radar, remote controls

VISIBLE LIGHT
Light bulbs

ULTRAVIOLET RADIATION
Sunbeds

X-RAYS
X-ray machines

GAMMA RADIATION
Nuclear radiation

The International Space Station

Constructed by, and on behalf of, the entire world, the International Space Station is the ultimate science fiction that's become reality: an epic structure orbiting above the Earth, and a fantastic viewing spot to get up-close-and-personal with the stars.

Look Up!

The International Space Station (ISS) is a feat of modern scientific engineering and technology. Approximately the size of an American football field (109m/357ft), the ISS has been home to 204 astronauts and scientists since the year 2000, when the first pieces of its construction were rocketed into space. Since then it has clocked up over 1.5 billion air miles and orbited the Earth more than 57, 631 times – up to 16 times a day!

Watch This Space

Weighing 419,459kg (924,739lb) – heavier than 320 cars! – the ISS is a state-of-the-art space laboratory that is larger than a conventional five-bedroom house, with a living room, two bathrooms, a gym and a 360° bay window – though most conventional houses are not controlled by 52 super-data-crunching computers.

The ISS's construction (still very much a work-in-progress) was the result of 115 space flights, and it houses more than 13km (8 miles) of wire cables. All its power is supplied by 4,047 sq. m (1 acre) of solar panels.

STELLAR FACT

The ISS takes just 90 minutes to make a full circle around Earth. It travels about 7.6km (4.7 miles) per second, at 27,358km (17,000 miles) per hour, above the Earth's surface.

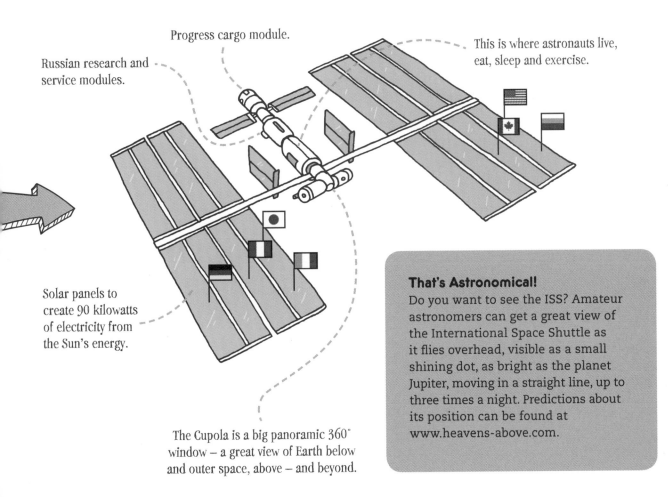

Progress cargo module.

Russian research and service modules.

This is where astronauts live, eat, sleep and exercise.

Solar panels to create 90 kilowatts of electricity from the Sun's energy.

The Cupola is a big panoramic 360° window – a great view of Earth below and outer space, above – and beyond.

That's Astronomical!
Do you want to see the ISS? Amateur astronomers can get a great view of the International Space Shuttle as it flies overhead, visible as a small shining dot, as bright as the planet Jupiter, moving in a straight line, up to three times a night. Predictions about its position can be found at www.heavens-above.com.

Probing the Void

Human beings love sending things into space. In the last 50 years mankind has launched thousands of satellites, men have walked on the Moon, rovers have reached Mars, and hundreds of space probes have ventured deep into the void of our Solar System, all in the name of progress and exploration. Let's have a look at the major space probes currently accelerating away from Earth.

Look Up!

You may not be able to spot these through your telescope but trust me, they're out there, dancing around the planets and hurtling through space like high-speed bullets. Since the 1950s dozens of space probes have been sent into space with important missions. Here are a few of the most important to whet your celestial appetite!

PIONEER 10 & 11

→ Launched *1972 and 1973*

The Pioneers were the first spacecraft sent to Jupiter and Saturn. En route Pioneer 10 travelled through the Asteroid Belt between Mars and Jupiter, making a flyby of the gas giant's famous Giant Red Spot. No longer sending back data, both probes will just continue their lonely journey into space forever.

VOYAGER 1

→ Launched *1977*

Current location: Interstellar space.

Deep-space probe Voyager 1, travelling at 62,763km (39,000 miles) per hour, is now the furthest object man has ever sent into space. In September 2013, NASA announced that Voyager 1 had entered interstellar space – the first manmade object to do so, at a distance of 18.7 billion km (11.6 billion miles) from the Sun.

GALILEO SPACE PROBE

Launched *1989*

Galileo was the first space probe to enter into orbit around Jupiter. Its mission was a long-term, detailed study of the gas giant and its moons. A smaller probe separated from the main spacecraft and entered Jupiter's atmosphere to make measurements there, the first probe to do so.

NEAR

Launched *1996*

The Near Earth Asteroid Rendezvous (NEAR) mission achieved history when it became the first space probe to land on an asteroid, 433 Eros, on 12 February 2001.

That's Astronomical!

Voyager 1 carries an audio-visual recording in the form of a gold-plated record that is meant both as a message to intelligent life and a symbolic time capsule. The record contains greetings in over 55 languages, pictures of Earth's life forms and recordings of music and sounds.

CASSINI-HUYGENS

Launched *1997*

The Cassini spacecraft has been orbiting around Saturn since 1 July 2004, sending back high-resolution images and other data about the planet, its rings and moons. The Huygens probe separated from the main Cassini spacecraft and became the first probe to land on the moon of another planet when it touched down on Saturn's largest moon, Titan.

A Star is Born

With over a hundred billion stars to choose from – and that's just in our own Milky Way – many astronomers have a favourite star. Which is your favourite?

Look Up!

Stars are like all living things. They are born, they live and they die. The Milky Way is home to many types of star, young and old, large and small, stable and violent.

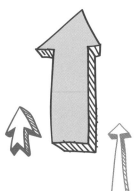

How A Star Is Born

★ Clouds of hydrogen gas and dust drift along for millions or even billions of years. When an event, such as the shockwave of a supernova or a collision of two clouds occurs, these clouds of dust and gas collapse into themselves due to gravity.

★ Once this happens, gravity then causes clumps to form and these draw gas inwards. The collapsing clump of stellar material begins to rotate and flatten into a disc of gas and dust.

★ The disc rotates faster and faster, pulling more material inwards and creating a hot, dense core. This is the beginning stage: now we have a protostar – this stage takes about 100,000 years to complete.

★ When the protostar becomes hot enough, hydrogen atoms begin to fuse, producing helium and energy. Astronomers call this the T Tauri stage.

★ After millions of years, a bipolar flow erupts from the protostar and blasts away any remaining gas and dust.

That's Astronomical!

You can tell how hot the surface temperature of a star is from its colour. The more massive the star, the hotter it is.

20,000°C blue-white
(the hottest stars, e.g. Rigel)

8,000°C white
(e.g. Sirius)

6,000°C Yellow
(e.g. Alpha Centauri A, our Sun)

4,500°C Orange
(e.g. Arcturus)

3,000°C Red (e.g. Betelgeuse, Proxima Centauri)

Our Expanding Universe

The Universe is getting bigger, and as it gets bigger the speed at which it is expanding is getting faster. Edwin Hubble was the first to prove that the Universe is getting bigger. This is bad news for astronomers ... or is it?

STELLAR FACT

In 2012, a team of astronomers used NASA's Spitzer Space Telescope to clock the expansion of the cosmos at an unbelievable 74km (46 miles) per second per megaparsec. A megaparsec is around 3 million light years.

Look Up!

It's important to understand that while the Universe is getting bigger, the galaxies, stars and planets we see in the sky are not really moving through space away from each other. Instead, what is happening is that the space between them is growing: think of it as a rubber band being stretched as two galaxies move apart. Because the Universe has no centre, as it expands, the galaxies get further away from each other, and the galaxies furthest from our Milky Way will move faster than those nearest to us. Everything is moving away from everything else.

Watch This Space

Another way to explain the expanding Universe is to imagine the Universe as a loaf of raisin-bread dough.

As the bread rises and expands, the raisins move further away from one another, but they are still stuck in the dough.

In the case of the Universe, there may be raisins out there that we can't see any more because they have moved away so fast that their light has never reached Earth. Thankfully, gravity is in control of things at the local level and keeps our raisins together.

From an astronomical perspective, when a star or galaxy is travelling away from Earth, the wavelength of light from this object is stretched out, making it look more red, this is known as the 'red-shift effect'. Edwin Hubble proved that the further a galaxy is from us, the more its light is red-shifted. This tells us that distant galaxies are moving away from us, and that the further a galaxy is, the faster it is moving away.

That's Astronomical!

In 1929 Edwin Hubble announced his discovery that the further away a galaxy is from another point in space, the faster it appears to recede as the Universe expands. This is known as Hubble's Law.

Supernovas

At the end of the life of the most massive stars, when they have become supergiants and exhausted their fuel, the star's core collapses in on itself, creating a supernova – an unthinkably huge and bright explosion, which releases a gigantic amount of energy. When a star dies in such spectacular fashion it leaves behind either a neutron star or a black hole in its place.

Look Up!

Stars do not stay the same forever, and their future depends on how much mass they have. A supernova can produce the same amount of energy in one second as an entire galaxy.

Stellar Fact

Astronomers living in ancient China recorded the first supernova some 2,000 years ago. They didn't understand what they were seeing and were convinced that the point of light was a new one.

Super Supernova Facts

☀ Stars about to go supernova change colour from red to blue due to their increasing temperature.

☀ Unlike a comet or commercial aeroplane, a supernova will stay in the same place once it has detonated.

☀ If you spot a supernova that isn't on record, you can report it to the IAU Central Bureau for Astronomical Telegrams.

That's Astronomical!

Scientists often rely on amateur astronomers to observe the skies while they're having a tea break. Ten-year-old Kathryn Aurora Gray of Fredericton, New Brunswick, Canada, discovered a supernova, making her the youngest person ever to find a stellar explosion. The Royal Astronomical Society of Canada announced the discovery – a magnitude 17 supernova in galaxy UGC 3378 in the constellation of Camelopardalis – on 2 January 2011.

TYPES OF STAR

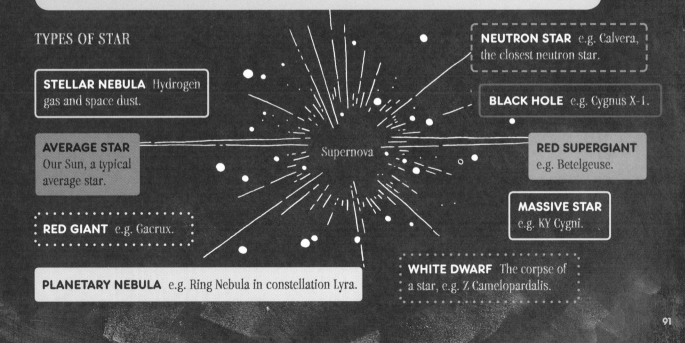

NEUTRON STAR e.g. Calvera, the closest neutron star.

STELLAR NEBULA Hydrogen gas and space dust.

BLACK HOLE e.g. Cygnus X-1.

AVERAGE STAR Our Sun, a typical average star.

Supernova

RED SUPERGIANT e.g. Betelgeuse.

MASSIVE STAR e.g. KY Cygni.

RED GIANT e.g. Gacrux.

WHITE DWARF The corpse of a star, e.g. Z Camelopardalis.

PLANETARY NEBULA e.g. Ring Nebula in constellation Lyra.

Extremely Large Telescopes of the Future

At the beginning of the 21st century, astronomers became even more determined to look further back in space and time. As you begin your adventures in astronomy and purchase your first telescope, if you have some spare money, you may want to think about buying one of these Extremely Large Telescopes ... the biggest ever created!

Look Up!

It's good to know that astronomers have a sense of humour. The newest and biggest telescopes in the world, known as ELTs, are exactly what they say on the tin – Extremely Large Telescopes.

An ELT is an observatory that features a reflecting telescope with an aperture of more than 20m (65½ft) in diameter, and is capable of scanning the skies, reflecting via wavelengths including ultraviolet (UV), visible, and near-infrared wavelengths. ELTs are designed to hunt for Earth-like planets around their parent stars as well as to peer deep into space.

That's Astronomical!

Mike Clements, a truck driver in Utah, USA, has set the record for building the largest ever amateur telescope – in his back garden. The telescope has a primary mirror of 178cm (70in), weighs 408kg (900lb) and stands 10.7m (35ft) tall. It took him 18 months to build.

European Extremely Large Telescope (EELT)

Watch This Space

There are many ground-breaking, ground-based telescopes all over the world. But let's have a look at the big three ELTs of the not-so-distant-future – telescopes that will be spinning the Universe on its head within the next 10 years.

 ## GIANT MAGELLAN TELESCOPE (GMT)

Due for completion 2020
Main mirror diameter 24.5m (80ft)
Location Las Campanas Observatory, Chile
Aims With seven of the largest and most precisely built telescope mirrors ever, the GMT will be used to study the early Universe and answer questions on dark matter, supermassive black holes and the nature of exoplanets.

EUROPEAN EXTREMELY LARGE TELESCOPE (EELT)

Due for completion early 2020s
Main mirror diameter 39.3m (129ft)
Location Cerro Armazones, Chile
Aims The EELT will gather 13 times more light than the largest optical telescopes that exist today and will study the first galaxies in the Universe, super-massive black holes and the nature of the Universe's dark matter.

THIRTY METER TELESCOPE (TMT)

Due for completion early 2020s
Location Mauna Kea, Hawaii
Main mirror diameter 30m (98ft)
Aims Started in 2007, this 56m (184ft) tall mega-scope will offer a resolution 12 times that of Hubble.

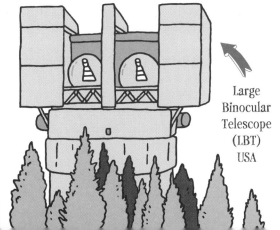

Large Binocular Telescope (LBT) USA

That's One Small Step

After the three-and-a-half-day flight it took to cover the 384,400km (238,862 mile) journey to their grey destination, astronauts Neil Armstrong and Buzz Aldrin left their command module pilot, Michael Collins, orbiting around the Moon ... and descended towards the rocky surface in their lunar module.

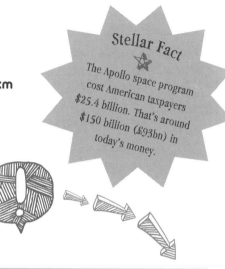

Look Up!

The first thing Aldrin and Armstrong did once they had piloted to a safe landing spot on the Moon, was to rehearse take-off. Just in case. Then they prepared themselves for the first spacewalk – a sight that would be watched by over half a billion people back home on Earth, and a record for the time.

In 1962, US President John F. Kennedy declared that by the end of the decade the country would put a man on the Moon ... and return him safely to Earth.

President Kennedy's desire to race to the Moon came at the height of the Space Race – a subplot between the two enemies of the Cold War, the USA and the Soviet Union.

In Greek mythology, Apollo was the son of Zeus, and was the god of light and the Sun.

Apollo 11 included a command module dubbed Columbia, piloted by Michael Collins, and a lunar lander called the Eagle, piloted by Neil Armstrong.

While Neil Armstrong took his first historic steps onto the lunar surface, Buzz Aldrin relieved himself inside a tube fitted in his spacesuit. The first urination on another planet, and live on TV!

If the launch from the Moon back to the command module had failed, space directors back in Houston, Earth had orders to cease and close down all communications and leave Armstrong and Aldrin to their deaths.

After eight days in total, the crew safely splashed down in the Pacific Ocean on 24 July 1969. They were immediately put into a three-week quarantine, for fear of unknown space pathogens.

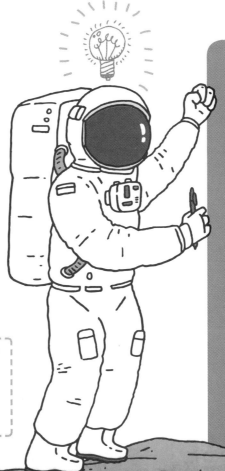

That's Astronomical!

After 21 hours on the Moon, Armstrong and Aldrin blasted off from the lunar surface to attach the top half of the returning lunar module, and Collins, onto the command module that would fire them back to Earth. But they almost didn't make it. Returning to the landing module after their spacewalk, Aldrin accidentally broke the switch used to activate the ascent engines. Fortunately he managed to activate the very important switch by stabbing the circuitry with his ballpoint pen; Aldrin carries the pen with him everywhere he goes.

Looking for Life on Mars

In 1976, Viking I was the first space probe to land successfully on, and send back data from, the famous red planet – and many astronomers' favourite view – **Mars. However, in 2012 the NASA rover Curiosity changed the way we look at the planet forever. The day the Curiosity rover started to beam back high-resolution images of Mars, 56.3 million km (35 million miles) away, is one all astronomers remember.**

Look Up!

After travelling for over nine months, the Curiosity rover – aboard the Mars Science Laboratory spacecraft (MSL) – finally entered the Mars atmosphere at 21,243km (13,200 miles) per hour, before spectacularly breaking away from the MSL and making a controlled landing, via an impressive 'sky crane', which lowered the rover safely onto the planet's surface.

Watch This Space

What do you know about Mars?

Mars Checklist

- ◯ Mars' mass is just over 10 per cent that of Earth.
- ◯ On Mars you can leap up three times higher in the air than you can on Earth.
- ◯ Mars has the highest mountain in the Solar System, 21km (13 miles) high.
- ◯ Mars has the largest dust storms in the Solar System.
- ◯ Mars has two satellites – Phobos and Deimos.
- ◯ Mars is named after the Roman god of war.

The first words uttered by a mission-control engineer when the rover landed were, 'We are wheels down on Mars. Oh my God.'

Curiosity's mission aims are to search for the basic ingredients essential for life – small concentrations of elements such as carbon, nitrogen, phosphorus, suphur and oxygen.

The rover is powered by nuclear power. Curiosity's generator has enough plutonium-238 dioxide to power itself for as long as 14 years, even though its mission will only last 23 months.

The project's final cost was a billion dollars more expensive than originally estimated, totalling $2.5 billion.

Curiosity was launched from Cape Canaveral, Florida, USA, on 26 November 2011.

The rover landed successfully on Aeolis Palus in Gale Crater on Mars on 6 August 2012.

ROVERS KEEP LEFT

Lasers fixed on top of the rover's 'head' can shoot at a distance of up to 7m (23ft) and vaporize anything in their path!

The MastCam is Curiosity's imaging tool. It captures high-resolution colour pictures and video of the Martian landscape.

Supermassive Black Holes

When a star burns out and supernovas, it either turns into a neutron star, or into something much more exciting and mysterious – a black hole.

Look Up!

At the centre of every galaxy lies a supermassive black hole – considered the last evolutionary stage of a star. Black holes are the cold, noisy and elusive leftovers of supernovas that are so dense that no matter, not even light, can escape from their super-powerful gravitational force. When light disappears from the edges of black holes in this way it is called an event horizon.

Watch This Space

Black holes may look like black holes, but don't let the name fool you: a black hole is anything but an empty hole. Instead, think of a black hole as a massive amount of material and matter packed into a tiny area. For example, think of a star ten times more massive than the Sun and imagine squeezing it into a sphere the size of a city such as London or New York. That's how dense a black hole is!

Black Hole Fact Checklist

O Black holes were predicted by Einstein's 1905 Theory of General Relativity, which outlined that when a massive star dies, it leaves behind a small, dense remnant core.

O If the star's core's mass is more than about three times the mass of the Sun, the force of gravity overwhelms all other forces and produces a black hole.

O If a black hole passes through a cloud of interstellar matter it will draw matter inwards; this is called accretion.

O The term 'black hole' was not coined until 1967.

O The nearest black hole to Earth is 1,600 light years away – that's far enough away not to panic.

O There is a black hole at the center of our Milky Way – but that's about 26,000 light years away.

That's Astronomical!

Black holes appear to exist on two different size scales:

☆ stellar mass black holes, around 10–24 times as massive as the Sun;

☆ supermassive black holes, which are billions of times as massive as the Sun.

Stellar Fact
✳ ✳ ✳

In 2003, astronomers using NASA's Chandra X-ray observatory, detected sound waves coming from a supermassive black hole 250 million light years away.

Fireworks in Space

Astronomers get to see fireworks all year round, in the form of meteors and meteor showers. Meteors are small, icy bodies that can be seen orbiting around the Solar System at great speeds. Comets are balls of rock and ice surrounded by a cloud of gas and dust that becomes visible when the comet approaches the Sun and begins to warm up.

Look Up!

Comets illuminate and streak through the sky thousands of times every night. As an astronomer, your chance of seeing one is incredibly high – just keep looking! The centre, or nucleus, of a comet is usually no greater than 16km (10 miles), which is very small compared to its tail, which can be up to 161 million km (100 million miles) long; radiation from the Sun pushes dust particles away from the centre of the comet to form the tail.

Watch This Space

Providing there is no light pollution, wherever you are on Earth, you will be able to see a few random meteors every night. Approximately 20 meteor showers occur each year that are visible to observers on Earth. Some of these showers have been around longer than 100 years. Here are some of the most spectacular.

- **Quadrantids** 2 January
- **Lyrids** 21 April
- **Eta Aquarids** 4–5 May
- **Delta Aquarids** 27–28 July
- **Perseids** 11–12 August
- **Orionids** 21 October
- **Leonids** 16 November
- **Geminids** 12–13 December
- **Ursids** 21–21 December

That's Astronomical!

Meteor showers get their names from
the constellation in which they are seen.
Perseids come from Perseus, hence the
name Perseids. Comets continuously
eject material with each passage
around the Sun and this replenishes
the shower with meteoroids.
Meteors are sometimes seen
with red, yellow or green trails.
The colours are caused by the
ionization of molecules, such
oxygen that appears green.

The impact theory is that a meteorite hitting Earth caused such devastation that the dinosaurs died out.

To Boldly Go ...
Where Next With Astronomy?

In an era of super-large space telescopes and planetary rovers on alien surfaces, we are undoubtedly set to learn much more about our Universe and our place in it over the next few decades.

Look Up!

As astronomy's breadth and knowledge has developed over the past century, so have the challenges and obstacles – both technological and political – that astronomers face in the future. Even though astronomers have only discovered and mapped a tiny percentage of space, this will not stop our desire to continue exploring the Universe. So, where next?

Watch This Space

The future of astronomy and space exploration is constantly evolving. With the construction the three largest ground-based telescopes currently happening (see page 93) and the development of the James Webb Space Telescope (JWST), what other grand plans do space agencies and astronomers have up their sleeves? Let's take a look.

1. MANNED MISSION TO MARS
Announced in 2012, the Mars One project is currently conducting a global search to find four suitable candidates for the first human mission to Mars, planned for 2023. Fancy it? It could be you!

2. THE MARS SPACESUIT

If you're going to travel to Mars on the Mars One project, be prepared to put on the Auoda.X spacesuit, the latest space fashion that will keep you alive. With all the cosmic radiation and toxic dust on Mars's surface, scientists have had to develop a new spacesuit for manned missions to Mars. This suit has been tested in a -110 degrees centigrade cryo chamber and on a five-day mock-Mars mission in a glacial ice cave!

3. THE SPACE DETECTION SATELLITE

Invented by the European Space Agency, this next-level satellite, also known as Gaia, launched on 19 December 2013. It is the most advanced satellite ever built and its goal is to make the largest, most precise 3-D map of space using its onboard 1 gigapixel digital camera. To put it another way, Gaia can see down to magnitude 20, which is 400,000 fainter than what we can see with the human eye!

4. ORION MULTI-PURPOSE CREW VEHICLE (MPCV)

The next-level vehicle of space exploration is NASA's awesome Orion MPCV. Built for crewed missions to Mars, the Moon and landing on asteroids, the Orion is NASA's Space Shuttle replacement. The first manned mission is due to take place in 2020.

Are We Alone in the Universe?

It's a question that has been bugging every astronomer since Galileo first used his telescope to look up at the sky over 400 years ago. Back then it was hard to imagine life beyond Earth – we were the centre of the Universe – but with the discovery of new planets with the potential to support life, the answer to this vital question is surely within reach in the next few decades.

'I'm sure the universe is full of intelligent life. It's just been too intelligent to come here.'

Arthur C. Clarke

Look Up!

In 1961, astrophysicist Frank Drake devised the Drake Equation in an attempt to estimate N – the number of potentially advanced civilizations in our Milky Way. Let's look at the Drake Equation in more detail:

$$N = R^* \; fp \; ne \; fl \; fi \; fc \; L$$

Or, put another way,

$$N = R^* \times fp \times ne \times fl \times fi \times fc \times L$$

N = number of civilizations in the Milky Way with which communication might be possible.

R^* = average rate of star formation in the Milky Way.

fp = fraction of those stars that have planets.

ne = average number of planets with the potential to support life per star that has planets.

fl = fraction of planets with the potential to support life that actually develop life.

fi = fraction of planets with life that actually develop intelligent life (civilizations).

fc = fraction of civilizations that develop a technology that produces detectable signs of their existence.

L = the length of time these civilizations continue to emit detectable signals into space.

Stellar Fact

According to Drake's Equation, there are about 10,000 technically advanced civilizations spread across the Milky Way.

That's Astronomical!

Since 1961, the Search for Extra-Terrestrial Intelligence project (SETI) has been looking for intelligent life in the Universe by listening out for radio waves sent by alien civilizations. Based in the USA and run by many of the country's top universities, SETI is a series of projects designed to search for intelligent life forms. No longer funded by the US government, privately funded companies are now in charge of SETI and are scanning the skies for signals from alien life.

Great Galloping Galaxies

Galaxies – one of the most thrilling sights to observe through a telescope – come in many shapes and sizes, with no two galaxies being the same. However, in the 1920s the Galaxy Morphological Classification, originated by Edwin Hubble, tried to classify and collate galaxies into a neat, organized system.

Look Up!
Seen through your telescope, galaxies can be divided into five distinct classifications.

1. SPIRAL GALAXIES

Our Milky Way, and the Andromeda system, are two great examples of spiral galaxies. As we are inside the Milky Way we cannot see its spiral shape, but look for the Whirlpool Galaxy (M51) and you'll see the famous pinwheel take shape.

2. BARRED SPIRAL GALAXIES

Far less numerous than spiral galaxies, astronomers are still unsure what makes a spiral galaxy produce a bar. The brightest barred galaxy we can see is M83 – take a look.

3. ELLIPTICAL GALAXIES

These galaxy types vary in size from dwarf to giant. The giants are the largest galaxies we can see – Messier 87, being one of the brightest and biggest.

4. IRREGULAR GALAXIES

Very faint and much smaller than spiral galaxies. Take a look at the Small Magellanic Cloud, the best example of an irregular galaxy.

5. PECULIAR GALAXIES

Try to spot M82 in the Ursa Major constellation. This bright example of a peculiar galaxy, as with all peculiars, shows signs of suffering from a mammoth disturbance such as a nearby star forming.

Watch This Space

If you are in the mood to go galaxy hunting, then grab your telescope and look at these beauties first.

NAME	CONSTELLATION	TYPE	DISTANCE FROM EARTH (MILLIONS OF LIGHT YEARS)
M31	Andromeda	Spiral	2.5
NGC5128	Centaurus	Elliptical	13
Large Magellanic Cloud	Dorado	Irregular	0.17
NGC253	Sculptor	Spiral	10
M33	Triangulum	Spiral	2.3
Small Magellanic Cloud	Tucana	Irregular	0.20
M81	Ursa Major	Spiral	7
M87	Virgo	Elliptical	40
M104	Virgo	Spiral	40

STELLAR FACT
Astronomers believe that there are probably more than 170 billion galaxies in the observable Universe, with each galaxy having, on average, 400 billion stars – if not more.

That's Astronomical!

At the heart, or nucleus, of every active galaxy lies a **quasar**. A quasar is a highly energetic object, burning a thousand times more brightly than the rest of the galaxies they inhabit. A quasar (or quasi-stellar radio source) occurs when gas near a supermassive black hole at the centre of a distant galaxy, enters the black hole at very high speed, but electromagnetic forces cause it to swirl around above the hole and blast off into space in the form of huge jets of energy.

Further Adventures Into The Unknown

Deep space exploration, and further adventures in astronomy, will continue to blow your mind away with the possibilities of all there is to learn about planet Earth and its place in the Universe. If you want to learn more, here are some ideas to get you started.

Websites

There are thousands of books on astronomy and stargazing, but these helpful websites will also help you point your binoculars in the right direction. These are my favourites, but you'll have fun finding your own too!

- www.skyandtelescope.com
- www.backyard-astro.com
- www.earthsky.org
- www.astronomynow.com
- www.nasa.gov
- www.galaxyzoo.org
- www.google.com/sky
- www.space.com
- www.hubblesite.org
- www.universetoday.com

Twitter

Twitter is the place to be to find up-to-date astronomy information, tweeted by experts such as former astronauts, cosmologists and other space-geeks. Here are my favourites.

- @NASA
- @ProfBrianCox
- @AstronomyMag
- @SETIinstitute
- @Spacedotcom
- @Chandraxray
- @NASA_Hubble
- @AsteroidWatch
- @skyandtelescope
- @discovery_space
- @astrospacenow

Equipment

Binoculars

Brilliant binoculars come in all shapes and sizes, depending upon your budget, and are a perfect way to introduce you to the heavens above if you are a beginner astronomer. Binoculars have a wide field of view, and can make zooming in on celestial objects near and far much more fun.

Telescopes

If your main passion in astronomy is exploring the finer details on planets or locating better images of distant galaxies, you will probably eventually need to buy a telescope, as binoculars simply won't have enough magnification. The best piece of advice on buying telescopes is to consult your friends and experts at your local astronomy club or contact a specialised retail outlet online or review-based website, such as www.astromart.com.

STELLAR FACT

✴ Binocular sizes are expressed with two numbers, such as 12 × 50. The first number refers to magnification. A pair of 12 × 50 binoculars will magnify the object you are looking at by 12. The second number relates to aperture, or diameter of the objective lenses in millimetres.

✴ Binoculars with larger apertures are better for astronomy, as they provide brighter images, but they are often heavier to hold.

✴ A pair of binoculars with a magnification of 7× to 12× and a large objective lens will show you planets in our Solar System, star clusters, nebulae and some galaxies.

✴ Any pair of binoculars to be used for astronomy with a magnification of more than 10× or 12× and or with objective lenses of 70mm or more will require a tripod.

✴ Zhumell Tachyon 25 × 100 binoculars are a good brand to start with.

Glossary

There are lots of big and awfully complex words to learn in astronomy. This is a small selection of some of the coolest words – they will make you sound clever at your next local astronomy club meeting.

Accretion disk A flat sheet of gas and dust surrounding a newborn star growing in size and attracting other stellar material.

Apparent magnitude The brightness of a star (or any celestial object) as seen from Earth.

Asteroids A class of small Solar System bodies in orbit around the Sun.

Astronomical unit (AU) The average distance between the Earth and the Sun – around 93 million miles (150 million kilometres). It takes a beam of light about 8.3 minutes to travel 1 AU.

Big Bang Cosmology's current best theory for the origin of the universe: an explosion of a tiny, ultra-hot spot of matter, around 13.8 billion years ago.

Binary star Two stars linked by mutual gravity and revolving around the same centre of mass.

Black hole A region round a small, but with incredible mass, collapsed star from which light cannot escape.

Comet A small body composed of ice and dust – with an elongated tail – which orbits the Sun on a lengthy orbit.

Constellation One of the 88 official patterns of stars into which the night sky is divided.

Corona The outermost part of the Sun's atmosphere, made up of gases.

Cosmic year The time it takes for the Sun to complete one revolution round the centre of the galaxy: approx. 225,000,000 years.

Cosmology The study of the Universe as a whole.

Dark adaptation The process by which the human eye increases sensitivity in darkness.

Dwarf star A small star in the hydrogen-burning phase of its life.

Electromagnetic spectrum The entire range of electromagnetic radiation from shortest to longest wavelengths. Gamma rays have the shortest wavelengths and radio wavelengths are the longest.

Exoplanet or Extrasolar planet A planet outside the Solar System.

Galaxy A system of stars, gas, dust, nebulae and other matter, bound by gravity and having a mass ranging from 100,000 to 10 trillion times that of the Sun. Most, but not all, galaxies are spiral.

Goldilocks zone The region around a star where a planet can maintain liquid water on its surface.

Infrared radiation Radiation with wavelengths longer than visible light.

Meridian An imaginary circle passing through the north and south poles, and through the observer's zenith, the point directly over their head.

Meteor The bright flash of light produced by a piece of space debris burning up as it enters the Earth's atmosphere at high speeds.

Milky Way A soft, glowing band of light encircling the sky, it is the disc of the spiral galaxy in which the Sun lies, seen from the inside.

Nebula A cloud of gas and dust in space.

Neutrino A tiny particle that is a by-product of nuclear fusion.

Neutron star Neutron stars send out radio emissions as they spin, creating a rapidly flashing source known as a pulsar.

Nova A star which flares up to many times its own brightness, remaining bright for a relatively short time before fading back down to its original brightness.

Objective The main light-gathering element in a telescope; it could be the lens or a mirror.

Orbit The path of a celestial object around another.

Photometry The measurement of the intensity of light.

Photon The smallest unit of light.

Proton A positively charged subatomic particle, the main ingredient of the atomic nucleus, alongside the neutron.

Protoplanet The earliest stages of the development of a planet.

Quantum The amount of energy possessed by one photon of light.

Quasar The core of a hugely powerful and active, but remote, galaxy.

Red giant A large reddish star in a late stage of its evolution.

Retrograde motion An orbital or rotational movement opposite to that of Earth's motion.

Scintillation The twinkling of a star due to the Earth's atmosphere.

Solar System Our Sun, and everything that orbits it.

Sunspot A highly magnetized dark spot on the Sun's surface, cooler than surrounding area.

Supernova A mammoth stellar explosion that involves the death of a massive star.

Terminator The boundary between the day and night hemispheres of the moon or a planet.

Ultraviolet (UV) The portion of the spectrum with wavelengths shorter than the bluest light visible.

White dwarf star A very small and dense star which has used up its nuclear energy and is in the very late stage of its evolution.

Zenith The observer's overhead point of, at an altitude of 90 degrees.

Zodiac A belt stretching around the sky, 8 degrees to either side of the ecliptic, in which the Sun and Moon can be found at any time.

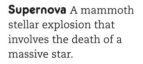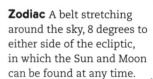

'Mortal as I am, I know that I am born for a day, but when I follow the serried multitude of the stars in their circular course, my feet no longer touch the Earth; I ascend to Zeus himself to feast me on ambrosia, the food of the gods.'

Ptolemy

Enjoyed this book? Collect the rest of the series!

Guaranteed 100% stodge-free and full of fun, the fact-tastic Cool series brings learning to life.

COOL Art

9781909396425

COOL MathS

9781907554841

COOL Science TRICKS

9781907554698

£9.99